Bringing Conservation to Cities

Lessons from Building the Detroit River International Wildlife Refuge

ECOVISION WORLD MONOGRAPH SERIES

Bringing Conservation to Cities

Lessons from Building the Detroit River International Wildlife Refuge

by
John H. Hartig

Ecovision World Monograph Series

Aquatic Ecosystem Health and Management Society
Burlington, Ontario, Canada

Front cover *(from top to bottom)*

- DTE Energy's Rouge Power Plant, River Rouge, Michigan – Nativescape, LLC
- Milliken State Park, Detroit, Michigan - SmithGroupJJR
- Milliken State Park, Detroit, Michigan - SmithGroupJJR
- Humbug Marsh, Trenton and Gibraltar, Michigan – Visual Image Productions
- *Main image:* Humbug Marsh, Trenton and Gibraltar, Michigan - U.S. Fish and Wildlife Service

Back cover: Milliken State Park, Detroit, Michigan - SmithGroupJJR

Cover design: J. Lorimer, AEHMS

ISBN: 978-0-9921007-4-2

First published 2014

Aquatic Ecosystem Health and Management Society
685 Inverary Road, Burlington, ON. Canada L7L 2L8

Printed by Volumes, Kitchener, Ontario, Canada.

Citation:

Hartig, J. H., 2014. *Bringing Conservation to Cities: Lessons from Building the Detroit River International Wildlife Refuge*. Ecovision World Monograph Series, Aquatic Ecosystem Health and Management Society, Burlington, ON, Canada.

Where There is No Vision, The People Perish

Proverbs 29:18

Table of Contents

Foreword

As a young boy growing up in the Detroit Metropolitan Area, I spent countless hours walking the shoreline of the Detroit River with my father, hunting in its many marshes, and fishing in its many spawning, feeding, and nursery areas. Those personal experiences with my father, a passionate outdoorsman, helped to shape my views on the importance of conservation and undoubtedly led me to use my position as a U.S. Congressman to promote responsible stewardship of the God-given natural resources entrusted to our care.

Growing up, my father would frequently remind me that we do not inherit this planet; we borrow it from future generations. This reminder has stayed with me and it has served as bedrock from which I have fought for conservation and environmental protection for my entire career. It is also the source of my passion for both building the Detroit River International Wildlife Refuge and making it accessible to the public. The story of our Detroit River International Wildlife Refuge is truly a remarkable one for many reasons:

- our conservation of continentally-significant fish and wildlife populations;
- our location in a major metropolitan area with nearly seven million people;
- partnerships with businesses, communities, nongovernmental organizations, and school children; and
- because we have improved outdoor recreational opportunities like fishing, hunting, wildlife observation, photography, environmental education, and interpretation.

That is why I'm so excited about this grand experiment of building an international wildlife refuge in the automobile capitals of the United States and Canada, and about how we are doing it in a fashion that improves public accessibility to natural resources, while engaging citizens in meaningful ways that lead to a sense of stewardship.

John Hartig has not only helped me build the Detroit River International Wildlife Refuge, but he has captured, in this book, how we have improved the environment and benefited our local communities and economies. Through the use of creative public-private partnerships, we are bringing conservation to cities and doing it in a fashion that makes nature part of everyday urban life. If we can bring conservation to the Detroit-Windsor Metropolitan Area and make nature part of everyday urban life in these automobile capitals, we can do it anywhere. I recommend this book to you, particularly its lessons learned, and hope that it will inspire you to bring conservation to your city.

Congressman John D. Dingell
United States House of Representatives

Editorial

The Aquatic Ecosystem Health and Management Society (AEHMS) publishes both a primary international journal, *Aquatic Ecosystem Health and Management,* and peer reviewed books under the banner of the Ecovision World Monograph Series.

Since 1995, the Ecovision series has published 23 books (see www.aehms.org; Table 1). The series focuses on the linkages between human society, ecology, economy, ecosystems and the environment, and publishes comprehensive and holistic treatments of a variety of topics dealing with whole ecosystems (interactions between air, water and land). This series merges the high quality of a journal with the comprehensive approach of a book.

The selection process for the Ecovsion books is stringent. To begin with, the proposal for a book is evaluated by the Ecovision Advisory committee in accordance with the aims and objectives of the Series. Following the approval of the proposal, the draft book is sent to two referees for peer review in addition to assessment by the Series editor. After the peer review, chapters are meticulously revised, condensed and checked. The revised version is then technically edited by the AEHMS before the book can be sent for printing.

Since its establishment in 1989, the AEHMS, via its journal and book Series, has encouraged and promoted integrated, ecosystemic, and holistic initiatives for the protection and conservation of global aquatic resources. I believe that John Hartig, in this monograph entitled, *Bringing Conservation to Cities: Lessons from Building the Detroit River International Wildlife Refuge,* has provided an outstanding example of fulfilling this mission. This is the second title which Dr. John Hartig chose to publish as a part of the Ecovision Series. In 2010, we published his book, *Burning Rivers: Revival of Four Urban-Industrial Rivers that Caught on Fire,* which won the Green Book Scientific Category Award in 2011.

Table 1. Books published under Ecovision World Monograph Series from 1995-2014.

Book title	Year
Aquatic Ecosystems of China: Environmental and toxicological assessment	1995
The Contaminants in the Nordic Ecosystem: Dynamics, Processes, and Fate	1995
Bioindicators of Environmental Health	1995
The Lake Huron Ecosystem: Ecology, Fisheries and the Management	1995
Phytoplankton Dynamics in the North American Great Lakes, Vol. 1: Lakes Ontario, Erie and St. Clair	1996
Developments and Progress in Sediment Quality Assessment: Rationale, Challenges, Techniques and Strategies	1996
The Top of the World Environmental Research: Mount Everest - Himalayan Ecosystem	1998
The State of Lake Erie Ecosystem (SOLE): Past Present and Future	1999
Aquatic Restoration in Canada	1999
Aquatic Ecosystems of Mexico: Scope & Status	2000
Phytoplankton Dynamics in the North American Great Lakes, Vol. 2: Lakes Superior, Michigan, North Channel, Georgian Bay and Lake Huron	2000
The Great Lakes of the World (GLOW): Food-web, Health & Integrity	2001
Ecology, culture and conservation of a protected area: Fathom Five National Marine Park, Canada	2001
The Gulf Ecosystem: Health and Sustainability	2002

Table 1. Cont'd.

Sediment Quality Assessment and Management: Insight and Progress	2003
State of Lake Ontario (SOLO): Past, Present and Future	2003
State of Lake Michigan (SOLM): Ecology, Health and Management	2005
Ecotoxicological Testing of Marine and Freshwater Ecosystems: Emerging Techniques, Trends, and Strategies	2005
Checking the Pulse of Lake Erie	2008
State of Lake Superior	2009
Burning Rivers: Revival of Four Urban-Industrial Rivers that Caught on Fire	2010

This current monograph tells the story of how innovative partnerships are making nature part of everyday urban life in the automobile capitals of the U.S. and Canada in an effort to inspire and develop the next generation of conservationists in urban areas, because that is where most people on our planet live. First, it provides a practical example of how sound science is being used as the foundation for ecosystem-based management of the Detroit River International Wildlife Refuge. Second, it is an excellent example of strengthening the science-policy linkage. Finally, I believe that this book provides unique insights and important lessons learned in both scientifically defining the concept of aquatic ecosystem health and in testing and evaluating the efficacy of ecosystem-based approaches to management of aquatic ecosystems on a global scale.

For those interested in ecosystem-based management of aquatic ecosystems, particularly in cities, and in bringing conservation to cities worldwide, this monograph is a must read. Aquatic Ecosystem Health and Management Society is honoured to include *Bringing Conservation to Cities: Lessons from Building the Detroit River International Wildlife Refuge* in its Ecovision World Monograph

Series. We welcome the new addition and are confident that the book will be a milestone ecosystem-based publication beneficial to researchers, managers and students dedicated to ecological protection and conservation.

I would sincerely like to thank Dr. John Hartig for choosing to publish in our Ecovision Series. We greatly appreciate the time consuming and thorough reviews conducted by Tom Edsall and John Gannon. Finally, thanks are also due to the AEHMS staff, namely Jennifer Lorimer, coordinator, Susan Blunt, technical editor and Robin Rozon and Lisa Elder for their assistance in the processing of this monograph. We are pleased to distribute and market the book with the Michigan State University Press, East Lansing, Michigan.

Mohiuddin Munawar, Series editor
Iftekhar Fatima Munawar, Managing editor

Ecovision World Monograph Series
Aquatic Ecosystem Health & Management Society
Burlington, ON. Canada
info@aehms.com

Dedication

To my colleagues and friends who I have worked with, side by side, for 15 years to create an exceptional, international, urban, wildlife refuge that: conserves biodiversity; reconnects people to world-class natural resources in their backyard; enhances quality of life; promotes ecotourism; increases community pride; helps make nature part of everyday urban life; and inspires that next generation of conservationists and sustainability entrepreneurs

Acknowledgements

Each of the many people and partners who have played instrumental roles in building the Detroit River International Wildlife Refuge has inspired me and helped develop this story of building an urban wildlife refuge in the industrial heartland. Clearly, this book would not have been possible without their contributions. I would like to specifically acknowledge:

- The individuals who helped craft the "Conservation Vision for the Lower Detroit River Ecosystem" in 2001;
- Congressman John Dingell, the late Peter Stroh, former Canadian Deputy Prime Minister Herb Gray, and Canadian Member of Parliament Jeff Watson for serving as high profile refuge champions;
- The staff of the U.S. Fish and Wildlife Refuge who helped guide the process and oversee refuge growth;
- The nonprofit friends group called the International Wildlife Refuge Alliance who has helped build the capacity to deliver conservation in this major urban area;
- The over 300 organizations in Canada and the United States who have worked in partnerships on refuge projects; and
- The numerous individuals, too many to count, who have led by example in respecting and loving this ecosystem as their home.

This book was written, in part, during writing residencies at Mesa Refuge in Point Reyes Station, California and Ragdale in Lake Forest,

Illinois. I gratefully acknowledge those unique opportunities to write without interruption in such inspirational locations.

I would also like to thank Dr. Mohi Munawar of the Ecovision World Monograph Series for his encouragement and support, Aquatic Ecosystem Health and Management Society staff for their significant contributions to the production process, and the independent reviewers whose comments improved this manuscript. Finally, I would like to especially thank my family for their continued support of my work and for their understanding of my passion for conservation and protecting the environment.

Prologue

In 1881 at the age of 17, Paul Kroegel moved from Germany to a homestead in Florida, USA overlooking Pelican Island in the Indian River. From his home, he could see the picturesque Pelican Island, a 2.02-ha (five-acre) mangrove haven where thousands of brown pelicans (*Pelecanus occidentalis)* and other water birds would roost and nest. It was an incredible sight of avian biodiversity and requisite nesting and roosting habitats comparable to many of the important birding areas featured through global ecotourism excursions today. He loved the birds of Pelican Island and became appalled over their wanton slaughter for feathers to appease the fleeting vanity of the fashion industry, particularly hats with exotic bird feathers. He went on to become a citizen activist for the conservation of the birds of Pelican Island. In one story he even put his life in jeopardy by positioning himself, all 1.7 m (5 feet 6 inches) tall, in his sailboat between the faster boats of gunners in an effort to protect his brown pelicans.

Kroegel patrolled the shores of Pelican Island with his shotgun trying to safeguard the nesting birds (U.S. Fish and Wildlife Service, 2011) (Figure 1). He spoke with anyone who would listen and eventually gained the ear of some prominent ornithologists who knew the young U.S. President Theodore Roosevelt who had a similar interest in birds. Kroegel's passionate advocacy led President Roosevelt on March 14, 1903 to establish Pelican Island as the first national wildlife refuge in the United States and the first unit of what eventually became the National Wildlife Refuge System (U.S. Fish and Wildlife Service, 2011). He was rewarded with becoming the first federal Warden of Pelican Island and is today considered a conservation hero. Now, more than a century later, the National Wildlife Refuge System has grown to become the largest network of public lands and waters dedicated to conservation of fish and wildlife in the world. This National Wildlife Refuge System is literally a conservation tapestry made up of over 550 national wildlife refuges that protect over 60.6 million ha (150 million acres) and that provide wildlife-compatible public uses like fishing, hunting, wildlife observation, photography, environmental education, and interpretation to 47 million annual visitors.

Figure 1. Paul Kroegel whose passion helped establish Pelican Island as the first national wildlife refuge in the United States (photo credit: U.S. Fish and Wildlife Service).

Aldo Leopold, another conservation hero, first became a forester, then a wildlife manager, and later a teacher who grew into a philosopher and gifted writer (Figure 2). He was a meticulous scientist, outstanding naturalist, and passionate conservationist. Today, he is considered by many the father of wildlife conservation in America and an influential leader of wilderness preservation. One of Leopold's greatest contributions was his book titled, *A Sand County Almanac,* through which he has inspired generations of people to have a deep sense of place as a basis of ethical human behavior (Leopold, 1949). Leopold first helped people understand that they were part of a community of interdependent parts and that the community included soils, water, plants, and animals. He felt that most people viewed the land as a commodity belonging to them and did not view the land as a community to which they belonged. He eloquently and persuasively got people to understand that "when we see land as a community to which we belong, we may begin to use it with love and respect." In essence Leopold called each of us to learn to live in a fashion that respects and loves the land, and all its inhabitants (Leopold, 1949).

Figure 2. Aldo Leopold – considered the father of wildlife conservation in America (photo credit: The Aldo Leopold Foundation).

With the use of the words "love" and "respect," Aldo Leopold intentionally integrated science and human behavior (Knight and Riedel, 2002). Leopold understood that conservation issues were not narrow and restricted, but multidimensional, requiring an integration of science, history, and culture to solve problems and achieve sustainable natural resource outcomes. His writing still rings true today because of how eloquently and persuasively he shared his personal struggles to integrate knowledge and make informed and ethical decisions about conservation of natural resources.

In tribute to his philosophy and contributions to conservation, the U.S. Fish and Wildlife Service established the Leopold Wetland Management District in 1993 and dedicated it to preserving, restoring, and enhancing wildlife habitat in Wisconsin, USA for the benefit of present and future generations. Today, the U.S. Fish and Wildlife Service (2011) uses a set of principles based on Leopold's philosophy to guide its conservation work within the National Wildlife Refuge System; the most significant of these states:

> *We are land stewards guided by Aldo Leopold's teachings that land is a community of life and that love and respect for the land is an extension of ethics. We seek to reflect that land ethic in our stewardship and to instill it in others.*

Rachel Carson, a third conservation hero, also has her roots in the U.S. Fish and Wildlife Service that manages the National Wildlife Refuge System. As a young child, she had a passion for exploring the forests and streams surrounding her hillside home near the Allegheny River in Pennsylvania, USA and for writing. She was first recognized for her writing at the age of 10 when she had a piece published in a children's magazine dedicated to the work of young writers. In 1925, she entered Pennsylvania College for Women as an English major determined to become a writer, but switched to biology part way through her studies.

She acquired an interest in oceans while working on a summer fellowship at the Woods Hole Oceanographic Institute in Woods Hole, Massachusetts, USA. Upon graduation from Pennsylvania College, she was awarded a graduate scholarship in biology at Johns Hopkins University in Maryland, an enormous accomplishment for a woman in 1929.

She went on to teach zoology at the University of Maryland, but her passions for writing and biology led to a job with the U.S. Bureau of Fisheries (now the U.S. Fish and Wildlife Service) in 1935. She served as a biologist with distinction, but was driven to translate her science in newspapers, magazines, and books. Her first book, titled *Under the Sea-Wind*, was published in 1941 and profiled her unique ability to explain complex scientific material relating to the natural world in clear poetic language for laypeople (Carson, 1941). Her second book, titled *The Sea Around Us*, was published in 1951 and remained on the New York Time's best-seller list for 81 weeks (Carson, 1951). As a result of the success of *The Sea Around Us*, she resigned her position with the U.S. Fish and Wildlife Service in 1952 to devote all of her time and energies to writing. Her next book, titled *The Edge of the Sea*, was published in 1955 and provided a new perspective on conservation to concerned environmentalists (Carson, 1955).

She is probably best known as the author of *Silent Spring*, a landmark exposition, published in 1962, that documented how pesticides like DDT were poisoning our rivers, lakes, oceans, and all life (Carson, 1962). *Silent Spring* helped catalyze the environmental movement, helped generate support for the establishment of Earth Day and banning of certain pesticides, and forever change society's approach to toxic substance management.

Today, she is considered the mother of the modern environmental movement, a world-renowned author who left a lasting legacy, and a true conservation hero. Her work helped catalyze the 1972 U.S. Clean Water Act, the 1972 Canada-U.S. Great Lakes Water Quality Agreement, the 1973 U.S. Endangered Species Act, and many other important environmental and natural resource laws (Figure 3). In honor of this extraordinary woman and her accomplishments in the environmental and conservation movements, the U.S. Fish and Wildlife Service named one of its national wildlife refuges near her summer home on the coast of Maine as the Rachel Carson National Wildlife Refuge in 1969.

All three of these conservation heroes would be so proud of the establishment of the Detroit River International Wildlife Refuge in 2001 as: the first international wildlife refuge in North America; an experiment in protecting continentally-significant natural resources in the industrial heartland; and a unique opportunity to bring nature and conservation into a major metropolitan area where millions of people live (Figure 4). Indeed, these three conservation heroes would be so pleased with the creation of this truly urban wildlife refuge that can make nature experiences part of everyday city life in an effort to help develop the next generation of conservationists (Table 1).

Figure 3. Rachel Carson – considered the mother of the modern environmental movement (photo credit: U.S. Fish and Wildlife Service).

Today, we need practical examples of where Kroegel's passion for protecting wildlife, Leopold's "land ethic," and Carson's advocacy for protecting the environment and conserving natural resources are being successfully applied in major urban areas, complete with measurable environmental, economic, and societal benefits. Large cities are typically where environmental, economic, and societal challenges are greatest. Detroit, Michigan and Windsor, Ontario are the automobile capitals of the United States and Canada, respectively. Together, they are considered part of the industrial heartland and "rust belt." During the 1960s, the Detroit River, that forms the international border between Canada and the United States, and that joins these to metropolitan areas economically, culturally, ecologically, and socially, was considered one of the most polluted rivers in North America.

Figure 4. An aerial photograph showing the Detroit River and western Lake Erie that are the heart of the Detroit, Michigan-Windsor, Ontario metropolitan area and that make up the Detroit River International Wildlife Refuge (photo credit: National Oceanic and Atmospheric Administration).

So, how could an international wildlife refuge be established in this major metropolitan region with nearly seven million people living in the watershed, industrial heartland with a long history of pollution, and "rust belt?" Isn't that a paradox?

It may seem like a paradox to build an international wildlife refuge in such a major metropolitan area that is part of the industrial heartland and the "rust belt," but it really isn't and you may be pleasantly surprised to learn why and how it is being done. That is the purpose and surprise of this book – to tell the compelling story of building North America's only international wildlife refuge in the industrial heartland of the United States and Canada, to share the important lessons learned about how it is being done in an effort to encourage more conservation initiatives in cities throughout the world, and to help inspire and develop the next generation of conservationists that must be developed with increasing frequency in major metropolitan areas because that is where most people on our planet live.

Table 1. Examples of exceptional natural resource attributes of the Detroit River International Wildlife Refuge.

Natural Resource Attribute	**Description**
Birds	Over 300 species of birds have been identified in the corridor
Flyways	Detroit River and western Lake Erie are situated at the intersection of the Atlantic and Mississippi Flyways
Waterfowl	30 species of waterfowl have been documented using the Detroit River; more than 300,000 diving ducks use the lower Detroit River as stopover habitat during spring migration

Table 1 Con't

Raptors	The lower Detroit River is one of the three best places to watch raptor migrations in the U.S.; 23 species of raptors migrate through the lower Detroit River; birders have seen over 100,000 raptors migrating in a single fall day
Important Bird Area	The lower Detroit River has been identified as an "Important Bird Area" by the National Audubon Society
Waterfowl hunting	In 2011, Ducks Unlimited identified Detroit as one of the top ten metropolitan areas for waterfowl hunting in the United States
Birding	Detroit River and western Lake Erie offer exceptional birding opportunities; a ByWays to FlyWays Bird Driving Tour Map features 27 unique birding sites in southwest Ontario and southeast Michigan
Fish	113 species of fish have been identified in the Detroit River
Fish spawning	Detroit River wetlands provide spawning areas for 26% of the fish species in the Great Lakes
Walleye	An estimated 10 million Walleye ascend the Detroit River from Lake Erie each spring to spawn, creating an internationally renowned sport fishery
Fishing	Detroit River and Lake Erie are considered the "Walleye Capital of the World;" major international fishing tournaments, sponsored by FLW Outdoors and other organizations, are held annually on the Detroit River and western Lake Erie offering prize money of as much as $1.5 million

Table 1. Cont'd

Biodiversity	The Detroit River and western Lake Erie have been recognized for their biodiversity in the: • the North American Waterfowl Management Plan (one of 34 waterfowl habitat areas of major concern in the U.S. and Canada); • the United Nations Convention on Biological Diversity (i.e. Detroit River and western Lake have identified as areas to receive biodiversity protection and conservation); • the Western Hemispheric Shorebird Reserve Network (i.e. marshes along the lower Detroit River and northeast Ohio have been identified a Regional Shorebird Reserve); and • the Biodiversity Investment Area Program of Environment Canada and U.S. Environmental Protection Agency (i.e. the Detroit River-Lake St. Clair ecosystem has been identified as one of 20 Biodiversity Investment Areas in the Great Lakes)
Wetlands of International Importance	Point Pelee National Park in Leamington, Ontario, Canada and Humbug Marsh in Trenton and Gibraltar, Michigan, USA have been identified as "Wetlands of International Importance" under the international Ramsar Convention
Heritage River Designation	The Detroit River is the first river in North America to receive both American Heritage River and Canadian Heritage River designations

Literature Cited

Carson, R.L., 1941. *Under the Sea Wind: A Naturalists Picture of Ocean Life*. Simon & Schuster, Inc., New York, NY.
Carson, R.L., 1951. *The Sea Around Us*. Oxford University Press, New York, NY.
Carson, R.L., 1955. *The Edge of the Sea*. Oxford University Press, New York, NY.
Carson, R.L., 1962. *Silent Spring*. Houghton Mifflin Company, Boston, MA.

Knight, R.L., Riedel, S., 2002. (Eds.). *Aldo Leopold and the Ecological Conscience*. Oxford University Press, New York, NY

Leopold, A., 1949. *A Sand County Almanac: And Sketches Here and There*. Oxford University Press, New York, NY

U.S. Fish and Wildlife Service, 2011. *Conserving the Future: Wildlife Refuges and the Next Generation*. Washington, D.C.

CHAPTER 1

From Resource Abuse to Recovery of Detroit River's Charismatic Megafauna

Most people have heard of the Laurentian Great Lakes. But not everyone recognizes how truly globally significant they are. Newcomers are always in awe of their size and sailors are frequently terrified of their power. The Great Lakes contain approximately 22,900 cubic kilometers (5,500 cubic miles) of water, representing nearly one-fifth of the standing freshwater on the Earth's surface. The Great Lakes drainage basin covers more land than England, Scotland, and Wales combined, and the lakes together have over 17,000 kilometers (10,000 miles) of shoreline. Think of this chain of lakes as a gigantic staircase, where the top step is Lake Superior and water descends down a series of lake steps to the Atlantic Ocean. There is little doubt as to why the Great Lakes are considered one of the great natural wonders of the world.

Situated at the heart of the Laurentian Great Lakes are the Detroit River and western Lake Erie. Lake Erie is one of the most intensively used large freshwater lakes in the world, serving multiple interests including fisheries, navigation, power generation, municipal waste assimilation, riparian, and recreation. It is the smallest of the Great Lakes by volume (483 km^3) and next to the smallest in surface area (25,700 km^2). It is also the southernmost Great Lake and the most biologically productive. This lake is naturally divided into the western, central, and eastern basins. The western basin is shallow (mean depth: 7 m), turbid because of high sediment loadings from the watershed, and mixes frequently as a result of winds.

The Detroit River is not a traditional river as most people understand it, but a 51.5-km (32-mile) connecting river system through which the entire upper Great Lakes (i.e. lakes Superior, Michigan, and Huron) flow to the lower Great Lakes (i.e. lakes Erie and Ontario). It provides 80% of the water inflow to Lake Erie (Bolsenga and Herdendorf, 1993).

This plethora of freshwater found in the Detroit River and western Lake Erie, and the exceptional fish, wildlife, and plant resources associated with it, provided sustenance to Native Americans for thousands of years (Hartig, 2003). From early French exploration, through European settlement, and finally through growth into a major urban industrial area, a number of "tipping points" occurred. Sociologists have used the concept of "tipping point" to define a moment in a series of events at which time significant, often momentous and irreversible reactions occur (Gladwell, 2000). A good way of thinking about it is the level at which the momentum for change becomes unstoppable. Environmental scientists refer to a "tipping point" as the point in time where there is urgent need to take action and that, if nothing is done, society could see irreversible damage or harm to an ecosystem (Bails et al., 2005). In the case of the Detroit River and western Lake Erie watershed, a number of environmental and natural resource "tipping points" have occurred over time that have resulted in considerable ecosystem damage and harm.

The first "tipping point" was the "Fur Trade Era" when Beaver were hunted to near extinction. French trappers and fur traders came in the late-1600s to conquer the region and reap its bounty. During this "Fur Trade Era" Beaver pelt hats were in high demand in Europe. To meet this high demand, Europeans eradicated much of the Beaver on their continent and then came to the Great Lakes, including the Detroit River-Lake Erie watershed, looking for Beaver.

Historians have noted that Detroit was literally created in response to European demand for hats made from Beaver pelts (Hartig, 2003). One of the best Beaver grounds in the whole Northwest Territory was Michigan's Lower Peninsula, especially the area between Lake Erie and Saginaw Bay (Publishers, 1980). A French officer named Antoine de la Mothe Cadillac came to the region we now know as Detroit in 1701 and established Fort Detroit to have a military presence at the "narrows of the river." This strategic placement of Fort Detroit was undertaken to both advance the fur trade and preserve it (Hartig, 2003).

Detroit went on to become a major center for collecting, processing, and exporting furs. It has been estimated that there were about 10 million Beaver in North America when Europeans first arrived (Dunbar and May, 1995). Europeans exported 50,000 skins annually until, by 1800, Beaver were near extinction. By the mid-1800s, fur trappers and traders had exhausted the region's supply of Beaver and the "Fur Trade Era" in Detroit was over. Merchants no longer depended on furs and the economy evolved to commerce, agriculture, and lumbering. Ceaseless slaughter of Beaver and loss of riparian forest habitat (i.e. through logging and clearing of land for agriculture) led to nearly wiping out Beaver in North America east of the Mississippi River by 1930.

The second "tipping point" occurred in the 1800s as rapid human population growth occurred in Detroit, resulting in substantial amounts of raw sewage being discharged into the Detroit River that was the very same place people were drawing their drinking water. This raw sewage caused waterborne disease epidemics. Detroit's first sewers, constructed in the early-1800s, were little more than open ditches through which raw sewage flowed into the Detroit River. As more and more raw sewage was discharged into the Detroit River, drinking water became contaminated, resulting in waterborne disease epidemics and a substantial human death toll. Examples of waterborne disease epidemics include:

- a two-week cholera epidemic occurred in Detroit in 1832, resulting 28 deaths; and
- a cholera epidemic occurred in 1834, killing 7% of Detroit's population (Hartig, 2003).

Both cholera and typhoid fever epidemics occurred in Detroit from 1832 through the early-1900s.

In response to the waterborne disease epidemics, Detroit began disinfecting its water supply in 1916 and moved its municipal water intake from along the mainland shoreline of the Detroit River to the head of the Detroit River at Belle Isle in 1932. As public wastewater and water treatment processes improved over time and water intake structures were relocated further upstream (including eventually southern Lake Huron), waterborne disease epidemics and associated mortality were eventually eliminated by the 1940s.

But alas, continued expansion of Detroit, particularly by industry, would eventually result in the third "tipping point." January of 1948 was the toughest month of a very harsh winter for waterfowl (Hartig and Stafford, 2003). First, it was a very cold winter with extensive ice cover and that meant that there were only a few pockets of open water where waterfowl could look for food. Second, it was right after World War II and substantial amounts of oil were still being dumped into the Rouge (a tributary of the Detroit River) and Detroit rivers. In fact, the U.S. Department of Health, Education, and Welfare (1962) estimated that during 1946-1948, approximately 22.44 million liters (5.93 million gallons) of oil and other petroleum products were being discharged into these rivers each year. To put this in perspective, this volume of oil being discharged was enough to pollute virtually the entire western basin of Lake Erie, including all Michigan, Ohio, and Ontario waters. That meant that oil covered the few remaining pockets of open water on the Detroit River and that waterfowl looking for pockets of open water landed in the oil. A few days later 11,000 waterfowl died from oil pollution (Cowles, 1975). Angry sportsmen from the Downriver area collected the oil-soaked carcasses of ducks and geese, put them in their pickup trucks, drove them to Lansing (the Capital of Michigan), and threw them on to the Capital lawn and sidewalk in protest. This single event is now credited with starting the industrial pollution control program of Michigan (Cowles, 1975).

During the 1960s, increasing inputs of phosphorus to Lake Erie led to excessive algal growth, oxygen depletion in the deeper waters of the lake, and massive fish kills. Picture front end loaders, the kind you see being used on highway construction projects, removing decomposing aquatic plants, algae, and dead fish from bathing beaches on Lake Erie.

Such widespread pollution of Lake Erie and its tributaries was prominently featured in national magazines, including Time magazine on August 20, 1965 (Time 1965). Time (1965) reported that:

> *Lake Erie is critically ill, and the symptoms are there for all to see. Beaches that once were gleaming with white sand are covered with smelly greenish slime. The lake's prize fish—Walleyes, Blue Pike, Yellow Perch and Whitefish—have all but disappeared, and the fishing fleets along with them.*

The national media coverage of the deterioration of Lake Erie, as evidenced by oxygen depletion and dead fish windrowing on the shorelines and beaches, led journalists to conclude that "Lake Erie is dead" (Leach, 1999). The resultant public outcry over the widespread pollution of Lake Erie became the fourth "tipping point."

Numerous citizens and stakeholder groups were outraged at the gross pollution and deplorable state of Lake Erie and its tributaries like the Detroit (Figure 1A and B) and Cuyahoga, and voiced their opposition at public water pollution hearings convened across the country, including ones in Detroit, Michigan and Buffalo, New York.

Figure 1. A: McClouth Steel Corporation discharging oil into the Detroit River in Trenton, Michigan, 1960 and B: Water pollution of the Detroit River coming from submerged outfall of the Detroit Wastewater Treatment Plant and from the Rouge River, 1966 (photo credits: A – City of Trenton; B – U.S. Environmental Protection Agency).

This public outcry over the pollution of Lake Erie and its tributaries went on to become a driving force in the establishment of the 1972 Clean Water Act and the 1972 U.S.-Canada Great Lakes Water Quality Agreement that called for a comprehensive approach to controlling phosphorus inputs to the Great Lakes. It is also interesting to note that in 1964 a laboratory of the Federal Water Pollution Control Administration was opened on an island in the lower Detroit River (Grosse Ile, Michigan) to address the public outcry over the water pollution of Lake Erie and its tributaries, including the Detroit River. At the creation of the U.S. Environmental Protection Agency in 1970, this laboratory was transferred to the U.S. Environmental Protection Agency, renamed the Large Lakes Research Station, and is still active today.

But Detroit, like all major cities, had not fully learned its lessons on pollution prevention and would experience yet another "tipping point." In 1967, an inquisitive Norwegian doctoral student named Norvald Fimreite came to the University of Western Ontario to study contaminant impacts on wildlife. Fimreite had studied mercury contamination found in Japan and Sweden during the 1950s and early-1960s, and was concerned about a possible similar situation in the Great Lakes. He estimated that in the late-1960s approximately 227 g (0.5 pounds) of mercury was lost to the St. Clair River (the connecting channel linking Lake Huron to Lake St. Clair and the Detroit River) for every 908 kg (one ton) of chlorine produced at Dow Chemical of Canada's chlor-alkali plant in Sarnia, Ontario (Turney, 1971). Government scientists went on to discover that, in total, approximately 91 tonnes of mercury were discharged into the St. Clair River, with additional amounts of mercury discharged into the Detroit River from Wyandotte Chemical Company. Fimreite followed up by sampling fish from Lake St. Clair in 1970 and found mercury concentrations over four times the standard for safe human consumption.

This news of mercury contamination of the St. Clair River and downstream environments descended on the people of Michigan and Ontario "like a thunder clap" (Turney, 1971). Mercury had contaminated the entire system, including the St. Clair River, Lake St. Clair, Detroit River, and western Lake Erie. The entire fishery from the St. Clair River to Lake Erie had to be closed, including shutting down 40 small family fisheries harvesting $1-2 million worth of fish each year (Hartig, 1983). The news of this mercury contamination spread rapidly across North America and became the fifth "tipping point" known as the "Mercury Crisis of 1970."

Such a long history with "tipping points," and the resultant public awareness of and outcry over long-standing environmental and natural resource problems in this major urban-industrial area, spurred the enactment of many environmental laws. The enactment of these laws stimulated many pollution prevention and control programs, and resulted in a long history of Canada-U.S. cooperation in investigating and monitoring the Detroit River-western Lake Erie corridor. Indeed, the 1972 U.S. Clean Water Act, the 1973 U.S. Endangered Species Act, the 1970 Canada Water Act, and the 1972 U.S.-Canada Great Lakes Water Quality Agreement have over a 40-year history. Further, the Detroit River Remedial Action Plan and the Lake Erie Lakewide Management Plan have nearly a 30-year history of identifying and implementing actions to restore impaired beneficial uses. The Comprehensive Conservation Plan for the Detroit River International Wildlife Refuge guides conservation efforts. The Lake Erie Committee of the Great Lakes Fishery Commission oversees fishery management planning, including goals, objectives, and implementation strategies. Each of these initiatives calls for monitoring to track changes and measure progress toward goals and objectives. In addition, many nongovernmental organizations and conservation groups practice citizen science through bird and plant surveys.

Indeed, it is often said that "if you can't measure it, you cannot manage it." Monitoring is essential for effective, defensible management. This chapter will document the environmental and ecological changes in the Detroit River and western Lake Erie over the last 40 or more years, and identify remaining challenges, based on the many long-term monitoring, surveillance, and citizen science programs established as a result of the "tipping points" identified above. Further, this chapter will show how the ecological recovery of the Detroit River has laid the foundation for building and celebrating the Detroit River International Wildlife Refuge.

Canada-U.S. Assessment of State of the Strait

The State of the Strait (Detroit in French means "the strait") Conference is a Canada-U.S. conference held every two years that brings together government managers, researchers, students, members of environmental and conservation organizations, and concerned citizens to assess ecosystem status and provide advice to improve research, monitoring, and management programs for the Detroit River and western Lake Erie. The 2004 Conference focused on monitoring for sound management and recommended that a binational indicator report be prepared to improve accessibility, science translation, and communication of long-term trends (Eedy et al., 2005).

An indicator is a measurable feature or features that singly or in combination provides useful information about status, quality, or trends. Indicators can be used to quantify the status for a whole array of factors from the state of the economy to the environment. Indicators should quantify information to make their significance apparent and convey it in a meaningful way to policy- and decision-makers, and the general public.

As a result of the recommendation from the 2004 State of the Strait Conference and the need for communication of status and trends of key indicators, the Detroit River-Western Lake Erie Indicator Project was initiated in 2004. The purpose of this project was to:

- compile and interpret long-term databases for 50 indicators from the Detroit River and western Lake Erie;
- translate the information into understandable terms for policymakers and managers; and
- make these data and their trends readily available (www.stateofthestrait.org).

Nearly 50 organizations and over 75 scientists participated in this three-year, indicator project for Detroit River and western Lake Erie that compiled long-term trend data on these 50 indicators, interpreted the data, translated the science for policy-makers and the public, and helped prepare a comprehensive and integrative assessment of ecosystem health. The focus of the 2006 State of the Strait Conference was to review these available indicator trend data and to help make recommendations to management on next steps and research needs.

Environmental Improvements and Ecological Revival

Collectively, the many long-term monitoring programs undertaken on the Detroit River and western Lake Erie have documented substantial environmental improvements since the 1960s (Table 1; Hartig et al., 2007a; Hartig et al., 2009), including:

- Substantial reductions in oil discharges and spills have occurred, and winter duck kills due to oil pollution have been eliminated;
- Billions of dollars have been spent on municipal wastewater treatment and virtually all plants in Michigan and Ontario are achieving secondary treatment;
- There has been a 90% decline in phosphorus concentration and loading from the Detroit Wastewater Treatment Plant (largest in the U.S.);
- Between 1960 and 2005 there has been a 65% reduction in untreated combined sewer overflow volume from communities in southeast Michigan;
- Over 4,000 tonnes per day of chloride loadings to the Detroit River were eliminated due to industrial process changes between the 1960s and 1980s;
- Substantial reductions in contaminants in fish have occurred (50-85% decline in mercury contamination and 70-90% decline in DDT contamination), yet health advisories remain in effect;
- Substantial reductions in contaminants in Herring Gull (*Larus argentatus)* eggs (90% decline in DDE and 85% decline in PCBs);
- Nearly one million cubic meters of contaminated sediment have been remediated at a cost of over $154 million (U.S.); and
- PCBs levels have declined by approximately 50% and mercury levels have declined by about 70% in Lake Erie sediment since the 1970s.

The combined effect of these environmental improvements over the past 40 years has been a surprising ecological recovery in this region, including an increase in the populations of sentinel indicator species like Bald Eagles (*Haliaeetus leucocephalus*), Peregrine Falcons (*Falco columbarius*), Osprey (*Pandion haliaetus)*, Common Terns (*Sterna hirundo*), Lake Sturgeon (*Acipenser fulvescens*), Lake Whitefish (*Coregonus clupeaformis*), Burrowing Mayflies (*Hexagenia* spp.), Chironomids, and Beaver (*Castor Canadensis)* in large areas from which they had been extirpated or negatively impacted (Table 2; Hartig et al., 2007a; Hartig et al., 2009).

Table 1. Evidence of environmental improvement in the Detroit River and western Lake Erie.

Indicator	Environmental Improvement	Reference
Oil pollution in the Rouge and Detroit rivers	Between 1946-1948 and 1961 there was a 97.5% reduction in oil discharges; between 1961 and the 1990s and early-2000s there was another order of magnitude reduction, with the exception of 2002 and 2004 when major oil spills occurred	Hartig et al., (2007b)
Municipal wastewater treatment in Michigan	In the late-1960s only primary treatment was provided at Michigan wastewater treatments plants; by the mid-1980s virtually all plants were providing secondary treatment	Fugita et al., (2007)
Phosphorus discharges from Detroit Wastewater Treatment Plant	Between 1966 and 2003 there was a 90% reduction in phosphorus concentration and loading	Fugita (2007)
Combined sewer overflow treatment in southeast Michigan	In 1960 none of the 119.2 billion liters per year of combined sewer overflow volume was treated; between 1960 and 2005 there has been a 65% reduction in untreated combined sewer overflow volume	Fugita et al., (2007)
Industrial chloride loadings to the Detroit River	Chloride loadings decreased from 4,247 tonnes day^{-1} in 1964-1966 to zero in 1986, when the last of five industries ceased operations that discharged high concentrations of chloride	Hartig and Perschke (2007)
Contaminants in fish	Since 1970 there has been an 85% reduction in mercury in Lake St. Clair Walleye; since the late-1970s there has been a 50% reduction in mercury in Lake Erie Walleye; since the late-1970s total DDT concentrations in Rainbow Smelt and Walleye have declined by 70% and 90%, respectively	Hayton (2007); Whittle (2007)
Contaminants in birds	Since the late-1970s DDE concentrations in herring gull eggs from Fighting Island have declined by 90%; PCB concentrations in Herring Gull eggs from Fighting Island have declined by 85% since the late-1970s	Weseloh (2007)
Contaminated sediment remediation	Since 1993 nearly one million cubic meters of contaminated sediment have been remediated at a cost of over $154 million (U.S.)	Zarull and Hartig (2007)

Table 2. Evidence of ecological recovery in the Detroit River and western Lake Erie.

Indicator	Evidence of Ecological Recovery	Reference
Bald Eagle reproductive success	Nearly complete reproductive failure by the mid-1970s; Eagle production in the region has increased since then and has leveled off in Ontario and Michigan with Ohio reproduction much greater and continuing to grow; in 2013 there were 22 active Bald Eagle nests in the vicinity of the Detroit River International Wildlife Refuge after a 25-year absence	Best and Wilke (2007); U.S. Fish and Wildlife Service
Peregrine Falcon recovery	Falcon population in Michigan decimated in the 1950s; falcons reintroduced in Detroit in 1987; since the early-1990s falcon reproductive success has steadily increased to 10 young in 2005; falcon removed from Endangered Species list in 1999	Yerkey and Payne (2007)
Osprey	In 2009, a pair of Osprey built a nest in a cell phone tower adjacent to the Gibraltar Wetlands Unit of the Detroit River International Wildlife Refuge, representing the first time that Osprey have successfully nested in Wayne County since the 1890s	U.S. Fish and Wildlife Service
Common Tern	In 2012, two Common Terns fledged from restored habitat created on the eastern tip of Belle Isle for the first time since the 1960s	U.S. Fish and Wildlife Service
Beaver	Beaver were hunted to near extinction during the "fur trade era;" during the height of oil pollution in 1940s-1970s, Beaver could not have survived in the Detroit River because oiled fur becomes matted and they lose their ability to trap air and water to maintain body temperature; loss of riparian forest habitat was another contributing factor; in 2008, two Beavers built a lodge at DTE's Conner Creek Power Plant; in 2009, this pair produced at least two pups; Beaver have continued to be seen through 2012; Beaver have also been reported from Belle Isle in the Detroit River, the headwaters of the Rouge River, in the Rouge River near University of Michigan-Dearborn, at Cross-winds Marsh; and at DTE's Rouge Power Plant	U.S. Fish and Wildlife Service

Table 2. Cont'd

Lake Sturgeon population	Substantial decline in Sturgeon population between the late-1800s and early-1900s; no Sturgeon spawning recorded from 1970s to 1999; in 2001 Sturgeon reproduction was documented for the first time in 30 years	Manny and Boase (2007)
Lake Whitefish spawning	Substantial decline in Whitefish population between the late-1800s and early-1900s; in 2006 Whitefish spawning in the Detroit River was documented for the first time since 1916	Roseman and Manny (2007)
Mayfly abundance in western Lake Erie	Few Mayflies present in western Lake Erie between 1950s and 1992; Mayflies increased between 1997 and 2004 being "good" to "excellent" in the biological reference point since 2002, but generally has shown high variability	Schloesser and Krieger (2007)
Chironomid abundance and diversity in western Lake Erie	Density increased four-fold between 1930 and 1961; similar in 1982; densities declined and richness increased by 1993, indicative of improving water quality conditions	Ciborowski (2007a)
Chironomid mouthpart deformities in the Detroit River	At the mouth of the Detroit River incidence of deformities has declined from two-times base-line in 1982 to baseline in 1994	Ciborowski (2007a)
Oligochaete Abundance in Detroit River	1960s to early-1990s had up to 1 million worms/m^{-2}; densities have since declined by 80-90%, but the numbers suggest some locations (e.g. Trenton Channel) still fall in the "heavily polluted" category	Ciborowski (2007b)

In the early-1900s Bald Eagles were distributed throughout Michigan, but by the 1950s the Bald Eagle population had significantly declined due to organochlorine pesticide contamination, loss of habitat, and other species changes. Reproductive impairments in Bald Eagles reached a crisis point in the mid-1970s when only 38% of Michigan's Bald Eagles could successfully fledge young. Bald Eagle recovery efforts were catalyzed by the banning of DDT in Michigan in 1969 and the remainder of the U.S. in 1972, and by the passage of the Endangered Species Act in 1973. From 1961 to 1987 there were no Bald Eagles produced in Michigan, likely due to contaminants and the low number of breeding pairs. Since 1991, Bald Eagle fledging success has steadily increased, particularly in Ohio where over 60 young were fledged in 2006 (Best et al., 2007). Fledging success in Michigan and Ontario has also increased since 1990, but to a lesser extent than Ohio. In 2013, U.S. Fish and Wildlife Service identified 22 active Bald Eagle nests in Monroe and Wayne counties in the vicinity of the Detroit River International Wildlife Refuge.

Peregrine Falcons experienced a dramatic decline in the 1950s, mostly due to organochlorine pesticide pollution. DDT caused reproductive problems in Peregrine Falcons and other species. As noted above, DDT was banned in Michigan in 1969 and in the remainder of the U.S. in 1972. A Peregrine Falcon reintroduction program was initiated in Detroit in 1987 when one nesting pair was re-introduced. No young were produced for the next five years. Then the number of young produced in southeast Michigan gradually increased from none in 1992 to a peak of 10 in 2005 (Yerkey and Payne, 2007). In 1999, the U.S. Fish and Wildlife Service removed Peregrine Falcons from the list of federally endangered species. In 2010, Peregrine Falcons expanded their range by successfully nesting on the Canadian side of the Ambassador Bridge that links southeast Michigan to southwest Ontario.

In 2009, a pair of Osprey built a nest in a cell phone tower adjacent to the Gibraltar Wetlands Unit of the Detroit River International Wildlife Refuge, representing the first time that Osprey have successfully nested in Wayne County since the 1890s. Osprey, also known as "fish hawks," are one of the largest birds of prey in North America, with a nearly 1.8-m (six-foot) wingspan. They feed almost exclusively on fish and are considered a good indicator of aquatic ecosystem health. As with Bald Eagles and Peregrine Falcons, a dramatic decline of Osprey occurred throughout North America due to widespread use of DDT and other organochlorine pesticides that caused eggshell thinning.

Common Terns are colonial water birds that are considered "threatened" in Michigan. Since the early-1960s there has been a 96% decline in nesting pairs of Common Terns along the Detroit River. This population decline has been attributed to loss of habitat, excessive mammalian predation, effects of contaminants, and competition with abundant Ring-Billed Gulls. Between 2008 and

2010, 800 m^2 of sand and crushed limestone were placed on Detroit Water and Sewerage Department property on eastern Belle Isle in an effort to restore nesting habitat. During the early-1960s as many as 1,200 nesting pairs of Common Terns were found on Belle Isle. In 2012, two Common Terns were confirmed to have fledged from restored Belle Isle habitat for the first time since the 1960s.

In 1978 the Walleye (*Sander vitreus)* population of Lake Erie was considered to be in crisis by fishery managers. The cause was overfishing and pollution. The Walleye population improved following a complete ban on fishing, as a result of the "Mercury Crisis of 1970", and the subsequent implementation of harvest quotas once the ban was lifted. The Walleye population of Lake Erie has shown year-to-year variability since the early-1990s (Haas and Thomas, 2007). In 2012, fishery biologists estimated that 22.2 million Walleye (age 2 and older) were present in Lake Erie, resulting in a total harvest of 2.48 million Walleye through sport and commercial fishing (Wills et al., 2013). This 2012 Walleye population estimate is below the long-term, 30-year average, primarily due to food web changes resulting from exotic species, elevated nonpoint source loadings of nutrients (particularly phosphorus), and climate change. Despite this recent decline in the Walleye population, Lake Erie and Detroit River are still considered the "Walleye Capital of the World" and attract major international fishing tournaments.

In the 1800s, Lake Sturgeon were considered a delicacy because of the sought after flavor of their smoked flesh and their eggs as caviar. In 1880, Lakes Huron and St. Clair produced over 1.8×10^6 kg (four million pounds) of Lake Sturgeon. During the spawning period in June 1890, over 4,000 adult Lake Sturgeon were caught in Lake St. Clair and the Detroit River on setlines and in pond-nets. Populations plummeted during the early-1900s due to over harvesting, limited reproduction, destruction of spawning habitats, and water pollution. Today, there is no active commercial fishery for Lake Sturgeon in the Huron-Erie corridor. Sport fishing harvest of Lake Sturgeon is now restricted in the St. Clair River and Lake St. Clair, and no sturgeon may be possessed by anglers from Michigan or Ontario waters of the Detroit River.

From the 1970s to 1999 no Lake Sturgeon spawning was reported in the Detroit River, which at one time was one of the most productive sturgeon spawning grounds in the United States (Manny and Boase, 2007). In 2001, however, Lake Sturgeon spawning was documented near Zug Island in the Detroit River for the first time in over 20 years (Manny and Boase, 2007). Improvements in water quality as a result of pollution prevention and control programs helped lay the foundation for Lake Sturgeon to once again spawn in the Detroit River. Additional information on Lake Sturgeon recovery efforts is presented in Chapter 7.

During the late-19th and early-20th centuries, large numbers of Lake Whitefish entered the Detroit River each fall to spawn. Natural bedrock spawning grounds were destroyed and removed during the construction shipping channel in 1907-1916. By the 1960s and 1970s Lake Whitefish numbers in Lake Erie were at a historical low because of overexploitation, predation by and competition with invasive species, degradation of water quality and habitat, and the loss of *Diaporeia*, their major, nutrient-rich food source. Improvements in water quality due to pollution prevention and control programs during the 1970s-1990s resulted in more favorable conditions for whitefish. In 2005, U.S. Geological Survey and U.S. Fish and Wildlife Service scientists documented natural reproduction of Lake Whitefish in the lower Detroit River for the first time since 1916 (Roseman and Manny, 2007).

Burrowing Mayfly populations were extirpated from western Lake Erie in the 1940s and 1950s as a result of water pollution. Burrowing Mayfly nymphs first reappeared in sediments of western Lake Erie in 1992-1993 after an absence of 40 years (Schloesser and Krieger, 2007). They returned in response to improved water quality resulting from pollution prevention and control programs, along with changes in trophic status, ascribed to Zebra Mussels (*Dreissena polymorpha).* Mayfly nymph densities have increased in recent years to a level considered good for healthy aquatic ecosystems. It is interesting to note that fishery managers have documented an increase in Yellow Perch (*Perca flavescens*) abundance in western Lake Erie between the early-1990s and early-2000s, likely due to increased Mayfly abundance.

In 2008, it is interesting to note that a pair of North American Beaver (*Castor canadensis)* returned to the U.S. side of the Detroit River and built a lodge at DTE's Conner Creek Power Plant. In 2009, this pair produced at least two pups. As noted earlier, Beaver were first hunted to near extinction during the "fur trade era" and then extirpated by over-hunting and loss of riparian forest habitat from urban development. During the height of oil pollution during the

1940s-1970s, Beaver could not have survived in the Detroit River because oiled fur becomes matted and they lose their ability to trap air and water to maintain body temperature. As of 2013, Beaver have been reported from six locations in the Detroit River watershed. Although the return of few Beaver cannot be considered ecologically significant, it is of great public interest and provides evidence of potentially expanding their range.

Collectively, this evidence of the return of sentinel fish and wildlife populations in the Detroit River represents one of the single most remarkable ecological recovery stories in North America. Indeed, 40 years of pollution prevention and control undertaken in response to the Canada-U.S. Great Lakes Water Quality Agreement, U.S. Clean Water Act, Canada Water Act, U.S. Endangered Species Act, and more have resulted in a return of charismatic megafauna.

Environmental and Natural Resource Challenges

This ecological recovery is remarkable, but the long-standing monitoring and surveillance programs in this corridor also document many environmental and natural resource challenges. Six of the most pressing ones include: population growth, transportation expansion, and land use changes; nonpoint source pollution; toxic substances contamination; habitat loss and degradation; introduction of exotic species; and greenhouse gases and global warming (Table 3; Hartig et al., 2007a; Hartig et al., 2009).

Over the last 60 years, a substantial portion of Detroit's human population has migrated out from Detroit to the suburbs, causing urban sprawl. Limited transportation options and unsustainable land use practices have contributed to urban sprawl, increasing habitat loss and degradation, and nonpoint sources pollution. Long-term monitoring programs have documented an increase in chloride concentrations in water samples from the Monroe Water Intake in recent years, a sharp increase in dissolved reactive phosphorus in water samples of the Maumee River in recent years, and an increase in total phosphorus concentrations in water samples of the Union Water Intake in Kingsville, Ontario in recent years. Scientists believe that food web changes resulting from the introduction of Zebra and Quagga Mussels, and nonpoint source pollution, are causing these changes. Further, these food web changes and nonpoint source pollution are being implicated in the resurgence of algal blooms. For example, in 2011 Lake Erie experienced the largest harmful algal bloom in its recorded history, with a peak intensity over three times greater than any previously observed bloom (Michalak, 2013). This record algal bloom is believed to be the result of record breaking phosphorus loadings from agricultural runoff and very high precipitation during the spring of 2011.

Detroit River and western Lake Erie have lost to development 97% and 90% of their coastal wetlands, respectively. Despite habitat conservation efforts, there remains considerable concern over the continuous and incremental loss and degradation of habitat in southeast Michigan and southwest Ontario. Exotic species like Zebra and Quagga Mussels, Round Gobies, Spiny Amphipod, and viral hemorrhagic septicemia virus have continued to impact the corridor, resulting in many economic, ecological, and societal impacts. Finally, evidence of global warming can already be found in the decrease of average winter ice cover on Lake Erie. Less winter ice cover increases evapotranspiration that contributes to reductions in water levels. Scientists have predicted that over the next 70 years, Lake Erie water levels may drop by 1.5 meters, resulting in a 4% reduction in lake surface area and a 20% reduction in water volume. Such reductions would move the Lake Erie shoreline lakeward by distances of 1-6 kilometers (Tyson and Ciborowski, 2007).

Concluding Thoughts

It is indeed amazing to think that the Detroit River was at one time considered one of the most polluted rivers in North America (Figure 1A and B). Indeed, this Great Lakes connecting channel has undergone incredible ecological revival from a polluted "rust belt" river to a natural resource asset that improves quality of life and enhances community pride. Educators talk about measuring student achievement in two basic ways: against a numerical standard like 60%, 70%, 80%, or 90%; and against a starting benchmark to be able to measure how far they have come. These educators would clearly be pleased with how far the Detroit River has come relative to where it started. This not only gives much to be proud of for the people of the Detroit-Windsor Metropolitan Area, but it gives a shining example of urban water resource revival that serves as a beacon of hope throughout the world.

Table 3. Key environmental and natural resource challenges for the Detroit River and western Lake Erie.

Environmental and Natural Resource Challenge	Evidence of Deteriorating Conditions	Reference
Population growth, transportation expansion, and land use changes	Steady increase in human population within seven county region; Detroit's population grew steadily from 1900 until 1950, then decreased by 50% between 1950 and 2005	Liu and Rogers (2007a)
	There has been a substantial increase in the conversion of agricultural land in southeast Michigan to urban development; there has also been a decrease in housing density	Liu and Rogers (2007b)
	88% increase in people driving personal vehicles since 1960; 26,000 fewer people using mass transit since 1980; average travel time to work has increased by three minutes in the last 20 years	Evans (2007)
Nonpoint source pollution	Dissolved reactive phosphorous in the Maumee River has increased from the mid-1990s through the mid-2000s, possibly due in part to nonpoint source pollution	Richards (2007)
	Chloride concentrations in water samples from the Monroe Water Intake have been increasing between the late-1980s and the early-2000s	Hartig and Perschke (2007)
	Total phosphorus concentrations in water samples from the Kingsville Water Intake have been increasing from the 1990s and early-2000s, possible due in part to nonpoint source pollution	Howell and Nakamoto (2007)
	Algal blooms returned to western Lake Erie in the 1990s and early- and mid-2000s; it appears that these algal blooms are linked to nutrient loading, nutrient releases from Zebra Mussels, and selective feeding by Zebra Mussels	Bridgeman (2007)
	97% of the coastal wetlands on both sides of the Detroit River have been lost to development since 1815; wetlands serve as living filters for nonpoint source pollutants	Manny (2007)
	Detroit ranks sixth among the 25 U.S. cities most polluted with particulate matter; since 1999 the Detroit area has met the new particulate matter air standard only once, in 2004; the Detroit area in not considered in compliance with the National Ambient Air Quality Standards for particulate matter and ozone	Wahl and Cummins (2007)

Table 3. Con't

Toxic substances contamination	Although contaminant levels in fish, Herring Gull eggs, and sediment have declined, health advisories remain in effect on the consumption of many species of fish	Hartig et al. (2007)
	Contaminated sediment in the Trenton Channel of the Detroit River, the lower Rouge River, the lower River Raisin, and the lower Maumee River continues to contribute beneficial use impairments	Zarull and Hartig (2007)
	Analysis of addled Bald Eagle eggs from Ontario and Ohio shows that criteria for adverse effects are consistently exceeded by PCB and dieldrin, and often exceeded by DDE concentrations	Best and Wilke (2007)
Habitat loss and degradation	There has been a substantial increase in the conversion of undeveloped and agricultural land to urban development; southeast Michigan has over five million people	Liu and Rogers (2007b)
	97% of the coastal wetlands on both sides of the Detroit River have been lost to development since 1815	Manny (2007)
	The Lake Erie shoreline along two counties in Ohio were almost unaltered until the 1930s; there has been a dramatic increase in shoreline hardening through the 1990s; shoreline hardening in Ottawa and Lucas Counties has increased to 78% and 98%, respectively	Livchak and Mackey (2007)
Introduction of exotic species	Zebra and Quagga Mussels arrived in the mid- to late-1980s; western Lake Erie maximum in early-1990s	Ciborowski (2007c)
	Vegetation at Erie Marsh was reported as stable from the early-1900s to the 1970s; common reed was reported at low densities during the 1950s; aerial coverage increased from 5 to 132 ha between 1984 and 2003 (over a 26-fold increase)	Pearsall and Muller (2007)
Greenhouse gases and global warming	Michigan's carbon emissions have increased by 59.7 million metric tons between 1960 and 2001, representing a 46% increase	Washington (2007)
	General decreasing trend of winter ice cover on Lake Erie as measured by Annual Maximum Ice Cover between 1963 and 2005	Assel (2007)
	Lake Erie water levels vary on both seasonal and long-term scales; high water levels were recorded in the 1950s and 1980s-1990s, while low water periods have occurred in the 1930s and 1960s; currently, water levels are slightly below the long-term average; water levels are projected to decline 1-2 m over the next 70 years	Tyson and Ciborowski (2007)
	Substantial increase in turkey vultures, suggesting a possible northern range expansion due to global warming	Hall-Brooks et al. (2007); Cypher (2007)

It clearly must be recognized that more needs to be done to fully realize the long-term goals of restoring the physical, chemical, and biological integrity of the Detroit River and western Lake Erie. However, the progress achieved to date, the informed, engaged, and vocal stakeholders and nongovernmental organizations involved, and the broad-based desire to create a sense of place in these watersheds bodes well for further improvement and recovery.

Scientists and resource managers, however, are worried about societal complacency. Indeed, the return of algal blooms in western Lake Erie and the elevated nonpoint source phosphorus loadings, particularly from the Maumee River, are of great concern. Experts are concerned today that Lake Erie may be at another "tipping point."

The five "tipping points" identified earlier in this chapter resulted in considerable environmental damage and harm. As a society, we cannot afford to reach another "tipping point." We need to make sure that we sustain our research and monitoring programs, and ensure that they are closely coupled with management efforts, in the spirit of adaptive management that assesses, sets priorities, and takes action in an iterative fashion for continuous improvement. Further, we need to make sure that the resulting scientific information and knowledge are translated for decision-makers and the public, and that they are broadly disseminated. This knowledge must then catalyze pollution prevention and remedial programs focused on restoring these ecosystems using all available technologies and resources.

Looking ahead, we must make sure that a high priority is placed on monitoring and research to measure changes, quantify remedial program effectiveness, and evaluate predicted benefits, and to be able to make mid-course management corrections, as necessary, under initiatives like the U.S.-Canada Great Lakes Water Quality Agreement, the Canada-Ontario Agreement, the Great Lakes Restoration Initiative, the Great Lakes Regional Collaboration, and other ones. We also need to invest in capacity-building, sustain and indeed grow our nongovernmental organizations, and place a higher priority on education to help develop the next generation of environmentalists, conservationists, and sustainability entrepreneurs.

There is no doubt that this ecological recovery has enhanced outdoor recreation and ecotourism. As a result, the Detroit River is now a major source of community pride. For example, the Detroit River is the only river system in North America to receive both Canadian and American Heritage River designations, and the Detroit

River and western Lake Erie are now the only international wildlife refuge in North America (i.e. the Detroit River International Wildlife Refuge). Indeed, this remarkable ecological recovery of the Detroit River laid the ecological foundation for creating and celebrating the Detroit River International Wildlife Refuge that will be discussed in the next chapter.

Literature Cited

Assel, R., 2007. Lake Erie ice cover. In: J.H. Hartig, M.A. Zarull, J.J.H. Ciborowski, J.E.Gannon, E.Wilke, G. Norwood, A. Vincent (Eds.), *State of the Strait: Status and Trends of Key Indicators*, pp. 102-104. Great Lakes Institute for Environmental Research Occasional Publication No. 5, University of Windsor, Windsor Ontario, Canada. ISSN 1715-3980.

Bails, J., Beeton, A., Bulkley, J., DePhilip, M., Gannon, J., Murray, M., Regier, H., Scavia, D., 2005. Prescription for Great Lakes Ecosystem Protection and Restoration: Avoiding the Tipping Point of Irreversible Changes. http://www.healthylakes.org/site_upload/upload/prescriptionforgreatlakes.pdf (April 2013).

Best, D.A., Wilke, E. 2007. Bald Eagle reproductive success. In: J.H. Hartig, M.A. Zarull, J.J.H. Ciborowski, J.E.Gannon, E.Wilke, G. Norwood, A. Vincent (Eds.), *State of the Strait: Status and Trends of Key Indicators*, pp. 267-275. Great Lakes Institute for Environmental Research Occasional Publication No. 5, University of Windsor, Windsor Ontario, Canada. ISSN 1715-3980.

Bolsenga, S. J., Herdendorf, C. E., (Eds.). 1993. *Lake Erie and Lake St. Clair handbook*. Wayne State University Press, Detroit, Michigan, USA.

Bridgeman, T., 2007. Algal blooms in western Lake Erie. In: J.H. Hartig, M.A. Zarull, J.J.H. Ciborowski, J.E.Gannon, E.Wilke, G. Norwood, A. Vincent (Eds.), *State of the Strait: Status and Trends of Key Indicators*, pp. 151-155. Windsor: Great Lakes Institute for Environmental Research Occasional Publication No. 5, University of Windsor. ISSN 1715-3980.

Ciborowski, J., 2007a. Chironomid abundance and deformities. In: J.H. Hartig, M.A. Zarull, J.J.H. Ciborowski, J.E.Gannon, E.Wilke, G. Norwood, A. Vincent (Eds.), *State of the Strait: Status and Trends of Key Indicators*, pp. 193-198. Great Lakes Institute for Environmental Research Occasional Publication No. 5, University of Windsor, Windsor Ontario, Canada. ISSN 1715-3980.

Ciborowski, J., 2007b. Oligochaete densities and distribution. In: J.H. Hartig, M.A. Zarull, J.J.H. Ciborowski, J.E.Gannon, E.Wilke, G. Norwood, A. Vincent (Eds.), State of the Strait: Status and Trends of Key Indicators, pp. 199-204. Great Lakes Institute for Environmental Research Occasional Publication No. 5, University of Windsor, Windsor Ontario, Canada. ISSN 1715-3980.

Ciborowski, 2007c., Invasion of zebra mussels (*Dreissena polymorpha*) and quagga mussels (*Dreisenna bugensis*). In: J.H. Hartig, M.A. Zarull, J.J.H. Ciborowski, J.E.Gannon, E.Wilke, G. Norwood, A. Vincent (Eds.), *State of the Strait: Status and Trends of Key Indicators*, pp. 205-211. Great Lakes Institute for Environmental Research Occasional Publication No. 5, University of Windsor, Windsor Ontario, Canada. ISSN 1715-3980.

Cowles, G., 1975. Return of the river. Michigan Natural Resources. 44(1), 2-6.
Cypher, P., 2007. Fall raptor migration over Lake Erie Metropark. In: J.H. Hartig, M.A. Zarull, J.J.H. Ciborowski, J.E.Gannon, E.Wilke, G. Norwood, A. Vincent (Eds.), *State of the Strait: Status and Trends of Key Indicators*, pp. 259-262. Great Lakes Institute for Environmental Research Occasional Publication No. 5, University of Windsor, Windsor Ontario, Canada. ISSN 1715-3980.
Dunbar, W.F., May, G.S., 1995. *Michigan: A History of the Wolverine State. 3rd Edition*. Eerdmans Publ. Co., Grand Rapids, Michigan, USA.
Eedy, R., Hartig, J., Bristol, C., Coulter, M., Mabee, T., Ciborowski, J., (Eds.). 2005. State of the Strait: Monitoring for Sound Management. Great Lakes Institute for Environmental Research, Occasional Publication No. 4, University of Windsor, Windsor, Ontario, Canada.
Evans, J., 2007. Transportation trends in southeast Michigan In: J.H. Hartig, M.A. Zarull, J.J.H. Ciborowski, J.E.Gannon, E.Wilke, G. Norwood, A. Vincent (Eds.), *State of the Strait: Status and Trends of Key Indicators*, pp. 193–198. Windsor: Great Lakes Institute for Environmental Research Occasional Publication No. 5, University of Windsor. ISSN 1715-3980.
Fugita, G., 2007. Phosphorus discharges from Detroit Wastewater Treatment Plant. In: J.H. Hartig, M.A. Zarull, J.J.H. Ciborowski, J.E.Gannon, E.Wilke, G. Norwood, A. Vincent (Eds.), *State of the Strait: Status and Trends of Key Indicators*, pp. 294-296. Great Lakes Institute for Environmental Research Occasional Publication No. 5, University of Windsor, Windsor Ontario, Canada. ISSN 1715-3980.
Fugita, G., Sherrill, J. Hinshon, D., 2007. Combined sewer overflow controls in southeast Michigan. In: J.H. Hartig, M.A. Zarull, J.J.H. Ciborowski, J.E.Gannon, E.Wilke, G. Norwood, A. Vincent (Eds.), *State of the Strait: Status and Trends of Key Indicators*, pp. 297-304. Great Lakes Institute for Environmental Research Occasional Publication No. 5, University of Windsor, Windsor Ontario, Canada. ISSN 1715-3980.
Gladwell, M., 2000. *The Tipping Point: How Little Things Can Make a Big Difference.* Little, Brown and Company, New York, New York, USA.
Haas, R., Thomas, M., 2007. The Walleye population of Lake Erie In: J.H. Hartig, M.A. Zarull, J.J.H. Ciborowski, J.E.Gannon, E.Wilke, G. Norwood, A. Vincent (Eds.), *State of the Strait: Status and Trends of Key Indicators*, pp. 226-229. Great Lakes Institute for Environmental Research Occasional Publication No. 5, University of Windsor, Windsor Ontario, Canada. ISSN 1715-3980.
Hall-Brooks, B, Pettit, B. Sodergren, J., 2007. Fall raptor migration at Holiday Beach Conservation Area, Amherstburg, Ontario. In: J.H. Hartig, M.A. Zarull, J.J.H. Ciborowski, J.E.Gannon, E.Wilke, G. Norwood, A. Vincent (Eds.), *State of the Strait: Status and Trends of Key Indicators*, pp. 253-258. Great Lakes Institute for Environmental Research Occasional Publication No. 5, University of Windsor, Windsor Ontario, Canada. ISSN 1715-3980.
Hartig, J.H., 1983. Lake St. Clair: Since the "Mercury Crisis." Water Spectrum. 15,18-25.
Hartig, J.H., 2003. *Honoring Our Detroit River, Caring for Our Home*. Cranbrook Institute of Science, Bloomfield Hills, Michigan, USA.
Hartig, J.H. Perschke, L,. 2007. Chloride levels in western Lake Erie water sampled collected from the Monroe Water Intake. In: J.H. Hartig, M.A. Zarull, J.J.H. Ciborowski, J.E.Gannon, E.Wilke, G. Norwood, A. Vincent (Eds.), *State of the*

Strait: Status and Trends of Key Indicators, pp. 111-115. Great Lakes Institute for Environmental Research Occasional Publication No. 5, University of Windsor, Windsor Ontario, Canada. ISSN 1715-3980.

Hartig, J.H., Stafford, T., 2003. The Public Outcry over Oil Pollution of the Detroit River. In, Hartig, J.H., (Ed.), *Honoring Our Detroit River, Caring for Our Home*, pp. 69-78. Cranbrook Institute of Science, Bloomfield Hills, Michigan, USA.

Hartig, J.H., Zarull, M.A., Ciborowski, J.H.H., Gannon, J.E., Wilke, E., Norwood, G., Vincent, A., (Eds.), 2007a. *State of the Strait: Status and Trends of Key Indicators.* Great Lakes Institute for Environmental Research Occasional Publication No. 5, University of Windsor, Windsor Ontario, Canada. ISSN 1715-3980.

Hartig, J.H., Zarull, M.A., Ciborowski, J.J.H., Gannon, J.E., Wilke, E., Norwood, N., Vincent, A., 2009. Long-term ecosystem monitoring and assessment of the Detroit River and Western Lake Erie. Environmental Monitoring and Assessment. 158: 87-104.

Hayton, A., 2007. Mercury in Lake St. Clair Walleye. In: J.H. Hartig, M.A. Zarull, J.J.H. Ciborowski, J.E.Gannon, E.Wilke, G. Norwood, A. Vincent (Eds.), *State of the Strait: Status and Trends of Key Indicators*, pp. 143-145. Great Lakes Institute for Environmental Research Occasional Publication No. 5, University of Windsor, Windsor Ontario, Canada. ISSN 1715-3980.

Howell, T. Nakamoto, L,. 2007. Phosphorus concentrations in the western basin of Lake Erie. In: J.H. Hartig, M.A. Zarull, J.J.H. Ciborowski, J.E.Gannon, E.Wilke, G. Norwood, A. Vincent (Eds.), *State of the Strait: Status and Trends of Key Indicators*, pp. 105-110, Great Lakes Institute for Environmental Research Occasional Publication No. 5, University of Windsor, Windsor Ontario, Canada. ISSN 1715-3980.

Leach, J.H., 1999. Lake Erie: Passages Revisited. In: M. Munawar, T. Edsall, I.F. Munawar (Eds.), *State of Lake Erie: Past, Present and Future*, pp. 5-22, Backhuys Publishers, Leiden, The Netherlands.

Liu, X., Rogers, J., 2007a. Human population growth and distribution in southeast Michigan. In: J.H. Hartig, M.A. Zarull, J.J.H. Ciborowski, J.E.Gannon, E.Wilke, G. Norwood, A. Vincent (Eds.), *State of the Strait: Status and Trends of Key Indicators*, pp. 52-56. Great Lakes Institute for Environmental Research Occasional Publication No. 5, University of Windsor, Windsor Ontario, Canada. ISSN 1715-3980.

Liu, X. Rogers, J., 2007b. Land use change in southeast Michigan. In: J.H. Hartig, M.A. Zarull, J.J.H. Ciborowski, J.E.Gannon, E.Wilke, G. Norwood, A. Vincent (Eds.), *State of the Strait: Status and Trends of Key Indicators*, pp. 57-62. Great Lakes Institute for Environmental Research Occasional Publication No. 5, University of Windsor, Windsor Ontario, Canada. ISSN 1715-3980.

Livchak, C. Mackey, S.D., 2007. Lake Erie shoreline hardening in Lucas and Ottawa Counties, Ohio. In: J.H. Hartig, M.A. Zarull, J.J.H. Ciborowski, J.E. Gannon, E.Wilke, G. Norwood, A. Vincent (Eds.), *State of the Strait: Status and Trends of Key Indicators*, pp. 85-90, Great Lakes Institute for Environmental Research Occasional Publication No. 5, University of Windsor, Windsor Ontario, Canada. ISSN 1715-3980.

Michalak, A.M., Anderson, E.J., Beletsky, D., Boland, S., Bosch, N., Bridgeman, T.B., Chaffin, J.D., Cho, K., Confesor, R., Daloglu, I., DePinto, J.V., Evans, M.A., Fahnenstiel, G.L., He, L., Ho, J.C., Jenkins, L., Johengen, T.H., Kuo, K.C.,

LaPorte, E., Liu, X., , McWilliams, M.R., Moore, M.R., Posselt, D.J., Richards, R.P., Scavia, D., Steiner, A.L., Verhamme, E., Wright, D.M., Zagorski, M.A., 2013. Proceedings of the National Academy of Sciences of the United States of America. 2013 March 1-5. Record-setting algal bloom in Lake Erie caused by agricultural and meteorological trends consistent with expected future conditions.. http://www.pnas.org/content/early/2013/03/28/1216006110.full.pdf+html?sid=db09a6c5-3db5-45e1-9a7b-1b184219da7d (April 2013).

Manny, B.A., 2007. Detroit River coastal wetlands. In: J.H. Hartig, M.A. Zarull, J.J.H. Ciborowski, J.E. Gannon, E.Wilke, G. Norwood, A. Vincent (Eds.), *State of the Strait: Status and Trends of Key Indicators*, pp. 172-176. Great Lakes Institute for Environmental Research Occasional Publication No. 5, University of Windsor, Windsor Ontario, Canada. ISSN 1715-3980.

Manny, B.A., Boase, J., 2007. Lake Sturgeon population. In: J.H. Hartig, M.A. Zarull, J.J.H. Ciborowski, J.E. Gannon, E.Wilke, G. Norwood, A. Vincent (Eds.), *State of the Strait: Status and Trends of Key Indicators*, pp. 221-225. Great Lakes Institute for Environmental Research Occasional Publication No. 5, University of Windsor, Windsor Ontario, Canada. ISSN 1715-3980.

Pearsall, D., Muller, B., 2007. Invasion of Erie Marsh Preserve by common reed (*Phragmites australis*) over the last 20 years. In: J.H. Hartig, M.A. Zarull, J.J.H. Ciborowski, J.E. Gannon, E.Wilke, G. Norwood, A. Vincent (Eds.), *State of the Strait: Status and Trends of Key Indicators*, pp. 177-180. Great Lakes Institute for Environmental Research Occasional Publication No. 5, University of Windsor, Windsor Ontario, Canada. ISSN 1715-3980.

Publishers, 1980. *The Saga of the Great Lakes*. Coles Publishing Company, Ltd. Toronto, Ontario, Canada.

Richards, R.P., 2007. Phosphorus loads and concentrations from the Maumee River. In: J.H. Hartig, M.A. Zarull, J.J.H. Ciborowski, J.E. Gannon, E.Wilke, G. Norwood, A. Vincent (Eds.), *State of the Strait: Status and Trends of Key Indicators*, pp. 68-74. Great Lakes Institute for Environmental Research Occasional Publication No. 5, University of Windsor, Windsor Ontario, Canada. ISSN 1715-3980.

Roseman, E. Manny B.A., 2007. Lake Whitefish spawning. In: J.H. Hartig, M.A. Zarull, J.J.H. Ciborowski, J.E. Gannon, E.Wilke, G. Norwood, A. Vincent (Eds.), *State of the Strait: Status and Trends of Key Indicators*, pp. 216-220. Great Lakes Institute for Environmental Research Occasional Publication No. 5, University of Windsor, Windsor Ontario, Canada. ISSN 1715-3980.

Schloesser, D. Krieger, K.A., 2007. Abundance of burrowing Mayflies in the western basin of Lake Erie. In: J.H. Hartig, M.A. Zarull, J.J.H. Ciborowski, J.E. Gannon, E.Wilke, G. Norwood, A. Vincent (Eds.), *State of the Strait: Status and Trends of Key Indicators*, pp. 189-192, Great Lakes Institute for Environmental Research Occasional Publication No. 5, University of Windsor, Windsor Ontario, Canada. ISSN 1715-3980.

Time, 1965. Time for transfusion. Vol. 86, No.8, August 20, 1965. www.time.com/time/magazine/article/0,9171,841998-1,00.html (April 2013).

Turney, W.G., 1971. Mercury pollution: Michigan's action program. Water Pollution Control Federation 43 (7), 1427-1438.

Tyson, J. Ciborowski, J.H.H., 2007. Lake Erie water levels. In, Hartig, J.H., Zarull, M.A., Ciborowski, J.J.H., Gannon, J.E., Wilke, E., Norwood, G., Vincent, A., (Eds.), *State of the Strait: Status and Trends of Key Indicators*, pp. 92-98, Great

Lakes Institute for Environmental Research Occasional Publication No. 5, University of Windsor, Windsor Ontario, Canada. ISSN 1715-3980.

U.S. Department of Health, Education, and Welfare, 1962. Pollution of the navigable water of the Detroit River, Lake Erie and Their Tributaries within the State of Michigan. Detroit, Michigan, USA.

Wahl, R. Cummins, G., 2007. Asthma hospitalization rates in Wayne County, Michigan. In: J.H. Hartig, M.A. Zarull, J.J.H. Ciborowski, J.E. Gannon, E.Wilke, G. Norwood, A. Vincent (Eds.), *State of the Strait: Status and Trends of Key Indicators,* pp. 276-280. Great Lakes Institute for Environmental Research Occasional Publication No. 5, University of Windsor, Windsor Ontario, Canada. ISSN 1715-3980.

Washington, A., 2007. Michigan's carbon emissions. In, Hartig, J.H., Zarull, M.A., Ciborowski, J.J.H., Gannon, J.E., Wilke, E., Norwood, G., Vincent, A., (Eds.), *State of the Strait: Status and Trends of Key Indicators*, pp. 75-77. Great Lakes Institute for Environmental Research Occasional Publication No. 5, University of Windsor, Windsor Ontario, Canada. ISSN 1715-3980.

Weseloh, D.V.C., 2007. Contaminants in herring gull eggs In: J.H. Hartig, M.A. Zarull, J.J.H. Ciborowski, J.E. Gannon, E.Wilke, G. Norwood, A. Vincent (Eds.), *State of the Strait: Status and Trends of Key Indicators*, pp. 146-150. Great Lakes Institute for Environmental Research Occasional Publication No. 5, University of Windsor, Windsor Ontario, Canada. ISSN 1715-3980.

Whittle, D.M., 2007. Contaminants in western Lake Erie fish community. In: J.H. Hartig, M.A. Zarull, J.J.H. Ciborowski, J.E. Gannon, E. Wilke, G. Norwood, A. Vincent (Eds.), *State of the Strait: Status and Trends of Key Indicators*, pp. 136-142. Great Lakes Institute for Environmental Research Occasional Publication No. 5, University of Windsor, Windsor Ontario, Canada. ISSN 1715-3980.

Yerkey, J.M. Payne, T., 2007. Peregrine Falcon reproduction in southeast Michigan. In: J.H. Hartig, M.A. Zarull, J.J.H. Ciborowski, J.E. Gannon, E.Wilke, G. Norwood, A. Vincent (Eds.), *State of the Strait: Status and Trends of Key Indicators*, pp. 263-266. Great Lakes Institute for Environmental Research Occasional Publication No. 5, University of Windsor, Windsor Ontario, Canada. ISSN 1715-3980.

Zarull, M.A. Hartig, J.H., 2007. Contaminated sediment remediation. In: J.H. Hartig, M.A. Zarull, J.J.H. Ciborowski, J.E. Gannon, E.Wilke, G. Norwood, A. Vincent (Eds.), State of the Strait: Status and Trends of Key Indicators, pp. 305-308. Great Lakes Institute for Environmental Research Occasional Publication No. 5, University of Windsor, Windsor Ontario, Canada. ISSN 1715-3980.

CHAPTER 2

From Conservation Vision to Establishment of an International Wildlife Refuge

If you have ever planned a big project, you know how difficult the initial planning can be. People often call the initial project planning phase "fuzzy" because it lacks a clear and compelling vision, and common expectations. In the early stages of project planning it is not uncommon for some stakeholders to have one vision and one set of expectations, while others have completely different ones. A clear vision is not a vague wish or dream, but a picture so clear and strong that it will help make the desired outcome real. For that reason, vision has often been described as "hope with a blueprint" (Senge, 1990). It should also be noted that defining a clear and compelling vision and common expectations in the early phases of project planning, often prevents wasting effort and avoids possible disappointment at the end of a project.

Defining this need was strategically addressed in a multi-stakeholder process for U.S.-Canadian cooperation for conservation efforts in and along the Detroit River. The idea was to clearly define a vision, without ambiguity, so that U.S. and Canadian stakeholders were working toward the same purpose and had the same expectations. Admittedly, at first it was a little "fuzzy." U.S. and Canadian scientists, natural resource managers, nongovernmental organization representatives, and concerned citizens knew that they needed to work together. However, to bring stakeholders from two countries together to reach agreement on a clear and shared vision is no simple task. Senge (1990) describes a shared vision as a force in people's hearts – a force of impressive power. At its simplest level, it is a picture that all stakeholders carry in their hearts and minds. It answers the question, "What do we want to create? Senge (1990) concluded that shared vision is a vehicle for building shared meaning.

A clear and shared vision creates a sense of commonality and gives coherence to diverse activities of stakeholder groups at all levels (Senge, 1990). It creates excitement and brings everyone together under a common purpose. If done correctly, it encourages new ways of thinking and acting, and encourages risk taking and experimentation. Senge (1990) argues that without a shared vision, a learning organization cannot occur.

Agreement on the Conservation Vision

There is no doubt that there are probably many ways to create a clear and shared vision, but the process of developing one for the Detroit River International Wildlife Refuge (DRIWR) was well received and noteworthy because of its success. In 2000, U.S. Congressman John Dingell, then Canadian Deputy Prime Minister Herb Gray, and corporate executive and Greater Detroit American Heritage River Initiative Chairman Peter Stroh brought a group of scientists, natural resource managers, and nongovernmental organization representatives together and charged them with defining a desired future state for the Detroit River ecosystem. The group included approximately 25 people from Canada and 25 people from the United States. Congressman Dingell, Deputy Prime Minister Gray, and Peter Stroh cared deeply about the region and recognized the importance of the Detroit River to the region's economy, communities, and quality of life. These three influential leaders used their stature to bring people together to do something exceptional for southeast Michigan and southwest Ontario. They recognized that this would have to come from the heart and be something that did not currently exist, but would have the potential to be transformational for the region.

The charge to this august workshop group was intentionally broad, including the possibility of building something, creating something, managing something, acquiring something, or doing something else. However, it had to be compelling and it had to result in a clear and shared vision. After two days of presentations and discussions, the group articulated a desired future state and conservation vision for the lower Detroit River ecosystem in a consensus document titled "A Conservation Vision for the Lower Detroit River Ecosystem" (Metropolitan Affairs Coalition, 2000). It was officially signed in December 2001 on behalf of Canada by then Canadian Deputy Prime Minister Herb Gray and then Canadian Member of Parliament Susan Whelan, and on behalf of the United States by Congressman John Dingell and then Greater Detroit American Heritage River Initiative Chairman Peter Stroh (Figure 2).

Figure 2. U.S. and Canadian dignitaries celebrate the signing of "Conservation Vision for the Lower Detroit River Ecosystem" in December, 2001 in Wyandotte, Michigan (Left to Right: Wyandotte Mayor Leonard Sabuda, Michigan Senator Chris Dingell, Greater Detroit American Heritage River Initiative Chairman Peter Stroh, Congressman John Dingell, Canadian Member of Parliament Susan Whelan, U.S. Senator Debbie Stabenow, Canadian Deputy Prime Minister Herb Gray, Member of Ontario Provincial Parliament Bruce Crozier, Michigan State Representative George Mans, U.S. Fish and Wildlife Service Regional Director Bill Hartwig, and Michigan State Senator Bill O'Neil (photo credit: Larry Caruso).

It was explicitly stated in the Conservation Vision document (Metropolitan Affairs Coalition, 2000) that the vision for the lower Detroit River ecosystem would:

- Provide strategic direction for habitat conservation programs in the lower Detroit River and support linkages with similar efforts in tributaries and their watersheds;
- Further binational coordination of efforts to conserve natural resources in this internationally significant region;

- Provide the rationale and direction for local conservation and land use planning initiatives, and illustrate their role in achieving the conservation vision; and
- Catalyze actions in both Canada and the United States to conserve and protect unique habitats and ecological functions for the benefit of present and future generations.

Recognizing the importance of the lower Detroit River ecosystem in sustaining quality of life, all U.S. and Canadian participants agreed to the following vision statement:

> *In ten years the lower Detroit River ecosystem will be an international conservation region where the health and diversity of wildlife and fish are sustained through protection of existing significant habitats and rehabilitation of degraded ones, and where the resulting ecological, recreational, economic, educational, and "quality of life" benefits are sustained for present and future generations.*

This conservation vision was also supported with the following elements that further defined the desired future state of the lower Detroit River (Metropolitan Affairs Coalition, 2000):

- Remaining marshes, coastal wetlands, islands, and natural shorelines are protected in perpetuity from developments;
- Degraded marsh, wetland, island, and shoreline habitats are rehabilitated, wherever and whenever possible, and protected in perpetuity;
- An International Wildlife Refuge has been established and is managed in a partnership consistent with the vision statement;
- The environment is clean and safe for all wildlife, fish, and other biota, including humans;
- Fish and wildlife communities are healthy, diverse, and self-sustaining;
- Levels of toxic substances do not threaten wildlife, fish, or human health;
- Economic development and redevelopment are well planned, aesthetically pleasing, and environmentally sustainable; and

- Public access and recreational and educational uses are seen as priorities for achieving "quality of life".

The concept of an international wildlife refuge had been born. As Senge (1990) noted, U.S. and Canadian stakeholders now had a clear picture of what they wanted. And it was a picture that could easily be carried in hearts and minds, and that could build shared meaning.

Achieving International Wildlife Refuge Legitimacy in the United States

Following binational agreement on the conservation vision in the 2000 workshop, U.S. Congressman John Dingell then introduced legislation in 2001, founded on the conservation vision, to establish the DRIWR. Congressman Dingell's intent was to codify an international wildlife refuge and ensure that it became a force in law and was carried in people's hearts and minds. The concept was easy to grasp – it would bring people together to create the first international wildlife refuge in North America and ensure that it would become a source of community pride in the industrial heartland, the automobile capitals of the United States and Canada, and the "Rust Belt." This legislation was signed into law by the President of the United States on December 21, 2001 (i.e. Detroit River International Wildlife Refuge Establishment Act; Public Law 107-91), establishing it as the only international wildlife refuge in North America and incorporating it into the U.S. National Wildlife Refuge System that includes over 550 refuges and over 60 million ha of land. A wildlife refuge is a geographical area where waters and lands are set aside to conserve fish, wildlife, and plants.

U.S. stakeholders believed that if an international wildlife refuge was created as part of the National Wildlife Refuge System, the structure, staff, and resources would follow. The goal was to create a refuge in 10 years. Under Congressman Dingell's leadership it happened in one year. The authorized Refuge boundary was eventually expanded to include islands, coastal wetlands, marshes, shoals, and riverfront lands along 76.8 km (48 miles) of shoreline of both the Detroit River and western Lake Erie (Figure 3). It also became one of only a few refuges situated in a major metropolitan area.

The U.S. Fish and Wildlife Service then developed a Comprehensive Conservation Plan for the DRIWR, consistent with National Wildlife Refuge System policy. This Comprehensive Conservation Plan was developed with broad stakeholder involvement, including participation from Canada (U.S. Fish and Wildlife Service, 2005). This gave further legitimacy to the DRIWR. The Comprehensive Conservation Plan adopted of the following vision statement:

> *The Detroit River International Wildlife Refuge, including the Detroit River and western Lake Erie basin, will be a conservation region where a clean environment fosters the health and diversity of wildlife, fish, and plant resources through protection, creation of new habitats, management, and restoration of natural communities and habitats on public and private lands. Through effective management and partnering, the Refuge will provide outstanding opportunities for "quality of life" benefits such as hunting, fishing, wildlife observation and environmental education, as well as ecological, economic, and cultural benefits, for present and future generations.*

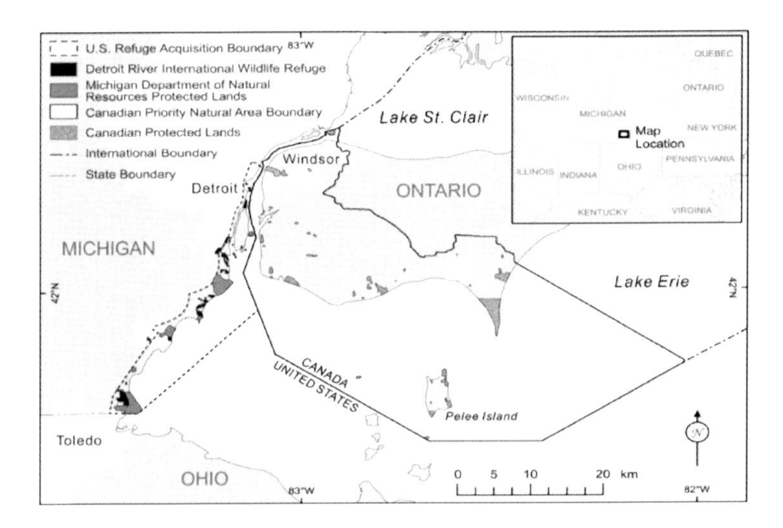

Figure 3. U.S. refuge acquisition boundary and Canadian Western Lake Erie Watersheds Priority Natural Area boundary that were used to build the DRIWR.

Of notable interest was the multi-stakeholder process used to build the DRIWR and foster a more sustainable future (U.S. Fish and Wildlife Service, 2005). Clearly, there is no single, best, multi-stakeholder process. Indeed, there are many successful ones. However, most successful approaches and processes recognize uncertainties and imperfect knowledge, integrate the environment with economic and social understanding, and practice adaptive management through an iterative decision-making process based on trial, monitoring, and feedback. Common elements of successful processes include: stakeholder involvement; leadership; information and interpretation; action planning within a strategic framework; human resource development; results and indicators; review and feedback; and stakeholder satisfaction. Figure 4 presents the general framework being followed for the DRIWR. The DRIWR has employed a multi-stakeholder process founded on the above elements, with an emphasis on implementing projects and taking actions that make progress toward the consensus vision and a sustainable future.

The refuge's Comprehensive Conservation Plan outlined how the refuge would fulfill its legal purpose and contribute to the National Wildlife Refuge System's wildlife, habitat, and public use goals (U.S. Fish and Wildlife Service, 2005). The plan articulated management goals for the first 15 years and specified the objectives and strategies needed to accomplish specific goals, including partnerships, wildlife-dependent uses, public environmental awareness, watershed development, refuge outreach, heritage values, healthy fish and wildlife communities, reduced toxic substances, sustainable economic development, beneficial water uses, and conflicting use resolution.

A formal refuge acquisition boundary was established on the U.S. side of the Detroit River and western Lake Erie (Figure 3). The Establishment Act authorizes the U.S. Fish and Wildlife Service to cooperatively manage and purchase lands within this boundary to grow the refuge and achieve its purposes. It was explicitly stated that the refuge would be built and managed through partnerships, including cooperative agreements and other voluntary mechanisms (U.S. Fish and Wildlife Service, 2005). The refuge on the U.S. side grew from 121 ha (300 acres) in 2001 to over 2,307 ha (5,700 acres) in the first 12 years. The management target established in the Comprehensive Conservation Plan for the U.S. side was to acquire and cooperatively manage 4,859 ha (12,000 acres) in the first 15 years (U.S. Fish and Wildlife Service, 2005).

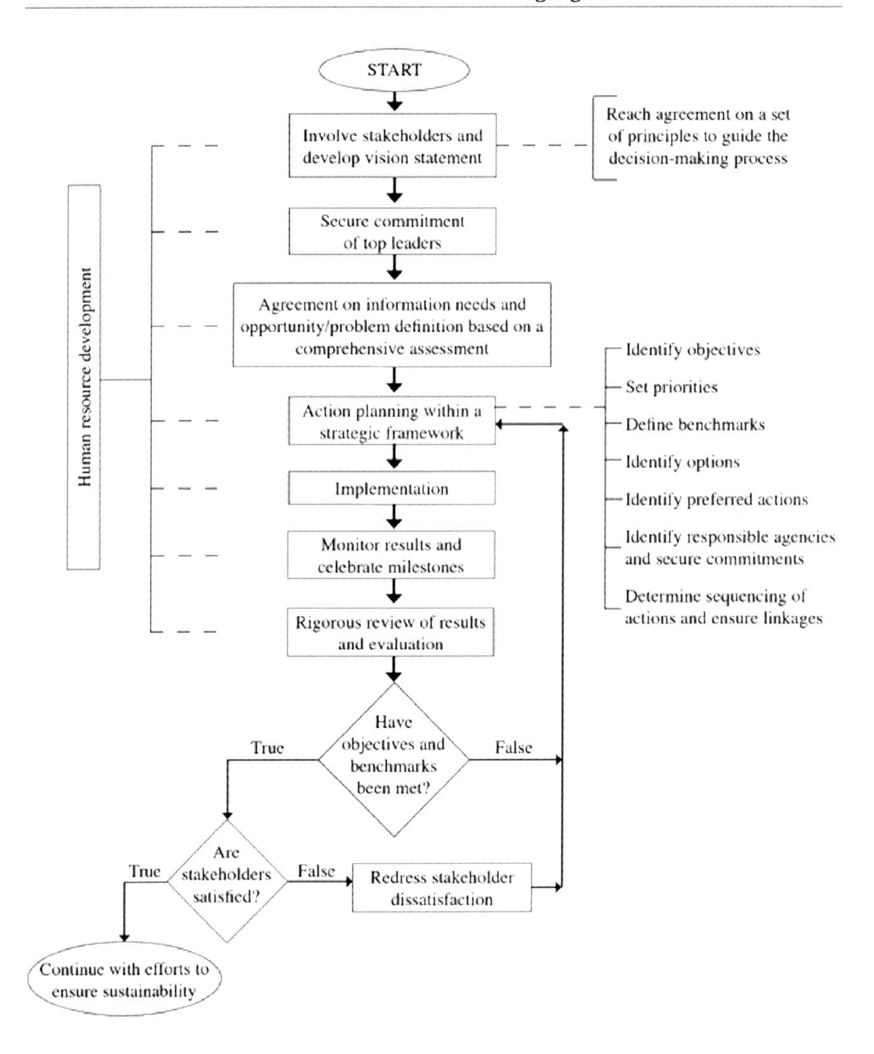

Figure 4. A multi-stakeholder process used to work collaboratively toward a sustainable future for the DRIWR, founded on adaptive planning and management.

Although the passage of the Establishment Act occurred relatively quickly, the process of operationalizing the refuge and building it to be sustainable took much longer. Many people understood that refuge staff and resources would not come immediately, but through a step-wise and incremental fashion. Looking back after the first 12 years, this was precisely what happened. A refuge manager was hired in 2004, an assistant refuge manager was hired one year later in 2005, a park ranger was hired in 2008, a refuge biologist was hired in 2011, and an administrative assistant was hired in 2014. Graduate students and contract employees were brought on board for specific projects as resources were secured. Partners were recruited on all projects to build the capacity to deliver results. Volunteers were recruited routinely to build capacity and a sense of ownership among citizens. Just like the refuge staff grew in a step-wise and incremental fashion, refuge projects started out small in an effort to build a record of success and build momentum.

In 2005, the U.S. Fish and Wildlife Service also established a Friends' Organization (i.e. an independent nongovernmental organization) called the International Wildlife Refuge Alliance to build the capacity of the refuge to work in partnerships and to advance its goals. Specifically, the International Wildlife Refuge Alliance's mission is:

> *to support the first International Wildlife Refuge in North America by working through partnerships to protect, conserve and manage the refuge's wildlife and habitats, and to create exceptional conservation, recreational and educational experiences to develop the next generation of conservation stewards.*

The International Wildlife Refuge Alliance hired its first staff person 2007. The importance and contributions of the International Wildlife Refuge Alliance cannot be emphasized enough. It helped substantially expand: outreach in the community; the refuge volunteer base; educational activities; stewardship activities; fund-raising; and advocacy. In essence, the Alliance creatively and effectively built the capacity of the U.S. Fish and Wildlife Service to deliver its mission in this major metropolitan region.

In 2013, the U.S. Fish and Wildlife Service signed a Memorandum of Understanding (MOU) with the Michigan Department of Natural Resources to cooperatively manage state and federal lands owned along the Detroit River and western Lake Erie in the spirit and intent of the 2001 Conservation Vision and the DRIWR. Michigan Department of Natural Resources and the U.S. Fish and Wildlife Service had a long history of monitoring and managing this corridor in the spirit of cooperative conservation and adaptive and ecosystem-based management. Through this MOU, Michigan Department of Natural Resources and the U.S. Fish and Wildlife Service maintain full management authority of their respective lands, but work collaboratively to create efficiencies and greater effectiveness on things like research, monitoring, conservation planning, restoration, and public use opportunities. As a result of the signing of this MOU, a U.S. registry of state and federal lands along the Detroit River and western Lake Erie was created showcasing how 2,342 ha (5,787 acres) of federal Refuge lands were being cooperatively managed with 3,196 ha (7,897 acres) of state land (i.e. Pointe Mouillee State Game Area – 1,635 ha or 4,040 acres, Pte. Aux Peaux – 89 ha or 220 acres, Erie State Game Area - 971 ha or 2,400 acres, Sterling State Park – 501 ha or 1,237 acres). That meant that 5,538 ha (13,684) acres of lands on the U.S. side were now being cooperatively managed in the spirit and intent of the 2001 Conservation Vision and the DRIWR for wildlife conservation and outdoor recreation.

Achieving International Wildlife Refuge Legitimacy in Canada

Following the signing of U.S. legislation for the Refuge in 2001, Canada responded by using a number of existing Canadian laws to work in a similar fashion. All U.S. and Canadian agencies agreed with the concept of the international wildlife refuge and pledged to work collaboratively to achieve the conservation vision. Environment Canada even participated in the drafting of the U.S. Comprehensive Conservation Plan and Environmental Assessment for the Detroit River International Wildlife Refuge (U.S. Fish and Wildlife Service, 2005) and noted that it was "working in partnership with the U.S. Fish and Wildlife Service and Canadian agencies to achieve a compatible, mutually shared, binational, focus for fish and wildlife habitat protection, conservation, and rehabilitation on the Canadian side of the Detroit River."

However, no full-time permanent staff was assigned to work on the Refuge in Canada. In practice, Canadian partner agencies worked opportunistically on projects like soft shoreline engineering, fish habitat restoration, and others. However, there was no institutional mechanism in the first ten years to add Canadian land to the Refuge. To address this deficiency and to help increase institutional collaboration, Canadian federal, provincial, and local partners signed a

Memorandum of Collaboration Agreement for the Western Lake Erie Watersheds Priority Natural Area in 2012. The Priority Natural Area begins at the head of the Detroit River near Peche Isle and includes all of the Detroit River watershed and the Ontario watershed of the western basin of Lake Erie, including Point Pelee National Park, Pelee Island, and the other offshore islands (Figure 2).

The Priority Natural Area initiative provided federal, provincial, and local partners with a mechanism to work more closely with U.S. partners and to complement progress being made in the U.S. on the DRIWR and other related initiatives such as the Detroit River Remedial Action Plan, the Detroit River Canadian Heritage River designation, and the Lake Erie Lakewide Management Plan. Federal, provincial, and local partners included Environment Canada, Fisheries and Oceans Canada, the Ontario Ministry of Natural Resources, the Nature Conservancy of Canada, Ducks Unlimited Canada, and the Essex Region Conservation Authority. It is anticipated that other partners, such as the agricultural community, municipalities, and others, will join the agreement as partners in the future.

"The Priority Natural Area initiative is the Canadian response to the Detroit River International Wildlife Refuge that reflects Canada's commitment to the Conservation Vision that was established in 2001," noted Canadian Member of Parliament Jeff Watson at the signing of the collaboration agreement in 2012. Canadian Member of Parliament Watson called the Agreement "historic", noting that the Priority Natural Area initiative would build on and complement existing programs and stimulate further conservation actions. The region's federal, provincial, local, and non-governmental partners had come together under the Western Lake Erie Watersheds Priority Natural Area initiative to better protect, restore, and manage the unique environmental values and health of the area of interest.

The vision agreed to by all parties was to:

Enhance collaboration and coordination of our resource management programs and projects, and engagement of the broader community as stewards in restoring, enhancing, and conserving the unique environmental values and health of the area of interest thereby making the region more attractive for people to live, work, and visit.

A steering committee made up of one senior representative of each party oversees the work of the initiative. Essex Region Conservation Authority provides local leadership and facilitation. The U.S. Refuge Manager of the DRIWR participates as a liaison representative.

Essex Region Conservation Authority, serving as the lead organization for the Priority Natural Area, then signed a Memorandum of Understanding with U.S. Fish and Wildlife Service to work collaboratively on transboundary conservation and outdoor recreational initiatives, and to create a Canadian registry of lands that would be cooperatively managed for conservation and outdoor recreation in the spirit of the 2001 U.S.-Canada Conservation Vision and the DRIWR. After the signing of this Memorandum of Understanding in 2013, 1,536 ha (3,797 acres) of Essex Region Conservation Authority lands and 396 ha (981 acres) of City of Windsor lands were added to the Canadian Registry of Lands.

Building the International Wildlife Refuge

Through considerable collaboration an international wildlife refuge had been created, founded on the common goals of conservation and sustainability. All stakeholders agreed to work through partnerships at all levels to move forward together toward this common vision. Through the institutional cooperation identified above, 7,471 ha (18,462 acres) were being managed as of 2013 for conservation and outdoor recreation in this southeast Michigan and southwest Ontario urban area that has nearly seven million people within a 45-minute drive. There clearly is, however, potential to grow the amount of land cooperatively managed for conservation and outdoor recreation to more than 10,117 ha (25,000 acres) by the 20th anniversary of Refuge establishment in 2021.

One simple thing that proved to be particularly beneficial was to translate the science of the environmental improvement and ecological revival that is summarized in Chapter 1. People were amazed to learn that the Detroit River revival was considered one of the most dramatic ecological recovery stories in North America. Further, this resonated throughout these "rust belt" cities and automobile capitals of Canada and the United States. People on both sides of the border were indeed so pleased to hear that the place they called home was improving, that additional improvements would follow, and that they lived in the watershed of the only international wildlife refuge in North America.

In the first ten years, to demonstrate binational cooperation on the refuge, a number of collaborative projects were also undertaken opportunistically between the United States and Canada to showcase the binational aspects of the refuge and to build trust and strengthen Canada-U.S. working relationships. Examples of cooperative projects are presented in Table 4. Strategically, this built a record of binational success while Canadian institutional arrangements were put in place to add land to the refuge on the Canadian side.

The institutional framework for the DRIWR has evolved over time as capacity was built, working relationships developed, and trust established. U.S. and Canadian laws are applied and enforced on each respective side of the international border. Governance is pursued collaboratively, attempting to achieve complementary and reinforcing management, policies, guidance, etc. This is particularly challenging in conservation initiatives, for example, because no single management agency is given clear authority for habitat management and responsibility is shared among numerous federal, state, provincial, and local agencies, and non-governmental organizations, corporations, and other private property owners. That is why it is often stated that "habitat has no home." To address this management challenge in the DRIWR, Canadian and U.S. agencies and organizations established a common vision and complementary and reinforcing goals and objectives. The reality is that any transboundary conservation initiative will always be challenging because of the international border, the variety of stakeholders, and complexity of decision-making.

The International Union for Conservation of Nature has been promoting knowledge and understanding of transboundary conservation (Erg et al., 2012). In its simplest form, transboundary conservation implies working across boundaries to achieve conservation objectives (Vasilijevic, 2012). One cannot talk about an international conservation initiative without talking about governance. Governance is about ensuring that an organization or initiative runs effectively and follows good practices in delivering its mission. Good governance frequently describes how institutions conduct public affairs and manage public resources. For governments, good governance often adheres to a number of key elements, including: accountability, transparency, efficiency and effectiveness, responsiveness, compelling vision of the future, and enforcement of laws.

Table 4. U.S.-Canada conservation projects undertaken for the DRIWR.

Project	Description
Conservation Vision for the Lower Detroit River Ecosystem	In 2001, stakeholders from the U.S. and Canada came together in a workshop to define a desired future state of the lower river (Metropolitan Affairs Coalition, 2001). This vision document, signed by senior representatives of the U.S. and Canada, called for the creation of an international wildlife refuge.
Comprehensive Conservation Plan for the Detroit River International Wildlife Refuge	Canadian agencies provided input on the development of the Refuge's Comprehensive Conservation Plan that guides management of the Refuge for the first 15 years.
ByWay to FlyWays Bird Driving Tour Map	This birding map highlights 27 exceptional birding locations in southeast Michigan and southwest Ontario. Examples include Point Pelee National Park in Canada, Lake Erie Metropark that is considered one of the three best places to watch Hawk migrations in North America, Rouge River Bird Observatory, and more. The map has been widely distributed as part of a binational ecotourism strategy. See Hartig et al., (2010) for more details.
Soft shoreline engineering	This project has catalyzed 53 projects that have enhanced shoreline habitat in the watershed of the Detroit River and western Lake Erie (see Chapter 5 for more details).
International Migratory Bird Day	International Migratory Bird Day (IMBD) is officially celebrated throughout the United States and Canada on the second Saturday in May. IMBD recognizes the movement of nearly 350 species of migratory birds from their wintering grounds in South and Central America, Mexico, and the Caribbean to nesting habitats in North America. IMBD has been jointly celebrated with public events in both countries on the same day.
State of the Strait Conferences	The State of the Strait Conference is held every two years to bring together government managers, researchers, environmental and conservation organizations, students and concerned citizens from Canada and the U.S. to assess ecosystem status and provide advice to improve research, monitoring, and management programs for the Detroit River and western Lake Erie. The conference alternates between Canada and the U.S. in the spirit of binational cooperation. See Hartig et al., (2009 and 2007) for more details.

Table 4 Con't.

Fighting Island Sturgeon Reef	In an effort to aid in the recovery of Lake Sturgeon, a threatened species in Ontario and Michigan, a surgeon spawning reef was constructed off the northeast corner of Fighting Island in 2008. Over 20 Canadian and U.S. partners participated. This project represented the first ever Canada-U.S. funded fish habitat restoration project in the Great Lakes. In the spring of 2009 natural reproduction was documented at the reef representing the first time in 30 years that Lake Sturgeon successfully spawned on the Canadian side of the Detroit River. See Chapter 7 for more details.
Common Tern Roundtable	In the spirit of adaptive management where population status is assessed, management priorities are established, and actions taken in an iterative fashion for continuous improvement, a Canada-U.S. roundtable was convened with Common Tern managers and Common Tern experts to establish an appropriate interim, quantitative target for the number of breeding pairs and their productivity that considers the population ecology of the species (Norwood et al., 2011). A population target of a mean of 780 breeding pairs across the region by 2020 was established. An interim goal for productivity is to reach at least 1.0 chick per nest across all colonies four out of every five years or a five-year mean of at least 1.0 chick per nest across all colonies. Management restoration actions and monitoring was also agreed to by partners consistent with Great Lakes basin-wide efforts.
Detroit River-Western Lake Erie Indicator Project	This binational project was a three-year study on long-term trends of key indicators of ecosystem heath of the Detroit River and western Lake Erie. Nearly 50 organizations and over 75 scientists participated in this three-year effort that compiled long-term trend data on 50 indicators, interpreted the data, translated the science for policy-makers and the public, and helped prepare a comprehensive and integrative assessment of ecosystem health. See Hartig et al., (2009 and 2007) for more information.
Raptor Migration Monitoring	Citizen science is used to systematically monitor raptor migrations across the Detroit River to track status and trends of raptor migrations. The monitoring program is undertaken at Holiday Beach in Amhurstburg, Ontario, Canada and at Lake Erie Metropark in Brownstown, Michigan, USA. Data are shared with natural resource managers and the public to support management efforts and educate and inform the public (Stein and Norwood, 2011).

Vasilijevic (2012) has shown that governance of transboundary conservation areas involves highly complex arrangements as these areas normally include and affect a wide variety of stakeholders, ranging from government agencies, nongovernmental organizations, local communities, the private sector, and even indigenous peoples. Shared governance, often called cooperative management, is "a partnership in which government agencies, local communities and resource users, nongovernmental organizations, and other stakeholders negotiate, as appropriate to each context, the authority and responsibility for management of a specific area or set of resources" (International Union for Conservation of Nature, World Commission on Protected Areas, 1997).

One useful way of looking at the level of international cooperation on the DRIWR is to utilize a numerical scale first developed by Zbicz (1999) and adapted by Sandwith et al., (2001) that ranks the level of conservation cooperation from none (Level 0) to full cooperation (Level 5; Table 5). Using this numerical scale, the current state of transboundary cooperation on the DRIWR is probably close to a Level 4 (coordination of planning). U.S. and Canadian stakeholders have cooperated on at least ten projects/activities that involve convening regular meetings, sharing information, coordinating planning, setting priorities, notification of emergencies, and, in some cases, undertaking shared projects from an ecosystem perspective.

Again, transboundary conservation projects in the refuge are undertaken opportunistically (Table 4). Other nationally-based projects are undertaken on both sides of the border, consistent with the common vision, goals, and objectives.

The institutional framework for the DRIWR is presented in Figure 5. On the U.S. side, leadership is provided by the U.S. Fish and Wildlife Service, under the auspices of the National Wildlife Refuge System. Cooperative management agreements are entered into with industries and other organizations to manage lands for conservation purposes. As noted above, a Memorandum of Understanding has also been signed between the U.S. Fish and Wildlife Service and the Michigan Department of Natural Resources to guide cooperative conservation within the U.S. portion of the Detroit River and western Lake Erie. The International Wildlife Refuge Alliance works with the refuge staff to build capacity, using a series of committees and special teams. The State of the Strait Steering Committee is binational in scope and make-up, whereas Detroit River Hawk Watch Advisory

Committee, Grosse Ile Nature and Land Conservancy, and the Cooperative Weed Management Area are focused just on U.S. initiatives. As noted above, on the Canadian side partners work through the Western Lake Erie Watersheds Priority Natural Area. U.S. Fish and Wildlife Service and Essex Region Conservation Authority have signed a Memorandum of Understanding to work collaboratively on transboundary conservation and outdoor recreational initiatives. Clearly, this institutional framework and transboundary cooperation will continue to evolve and improve consistent with an adaptive management philosophy.

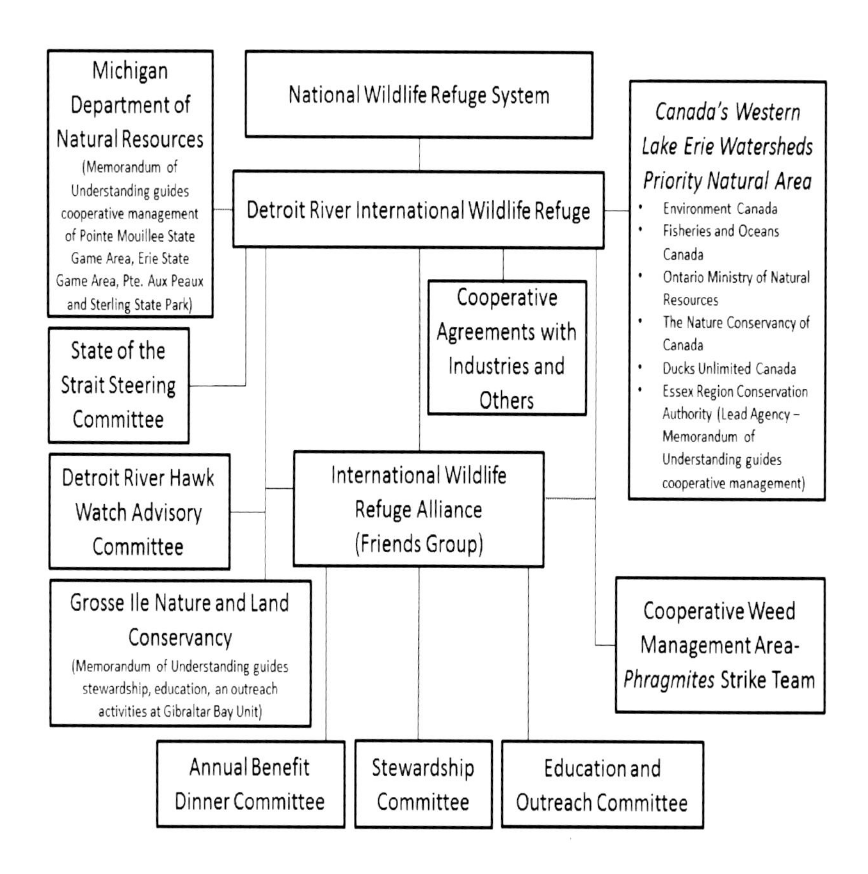

Figure 5. An institutional framework for managing the DRIWR.

Table 5. Levels of cooperation between internationally adjoining conservation areas (Zbicz, 1999; Sandwith et al., 2001).

Level of Cooperation	Characteristics
Level 0 – No Cooperation	• Staff from two conservation areas never communicate or meet • There is no sharing of information or cooperation on any specific issues
Level 1 – Communication	• There is some two-way communication between conservation areas • Meetings/communication takes place at least once a year • Information is sometimes shared • Notification of actions which may affect the other conservation area will sometimes take place
Level 2 – Consultation	• Communication is more frequent (at least two times per year) • Cooperation occurs on at least two different activities • The two sides usually share information • Notification of actions affecting the adjoining conservation area usually occurs
Level 3 – Collaboration	• Communication is frequent (at least every two months) • Meetings occur at least three times per year • The two conservation areas actively cooperate on at least four activities, sometimes coordinating their planning and consulting with the other conservation area before taking action
Level 4 – Coordination of planning	• The two conservation areas communicate often and coordinate actions in some areas, especially planning • The two conservation areas work together on at least five activities, holding regular meetings and notifying each other in case of emergency • Conservation areas usually coordinate their planning, often treating the whole area as a single ecological unit
Level 5 – Full cooperation	• Planning for the two conservation areas is fully integrated, and, if appropriate, ecosystem-based, with implied joint decision-making and common goals • Joint planning occurs, and, if the two share an ecosystem, the planning usually treats the two conservation areas as a whole • Joint management sometimes occurs, with cooperation on at least six activities • A joint committee exists for advising on transboundary cooperation

Concluding Thoughts

A clear and compelling vision was necessary to create the DRIWR and build shared meaning among stakeholders from both Canada and the United States. The 2001 "Conservation Vision for the Lower Detroit River Ecosystem," that was signed on behalf of Canada by then Canadian Deputy Prime Minister Herb Gray and Canadian Member of Parliament Susan Whelan, and on behalf of the United States by U.S. Congressman John Dingell and then U.S. Chairman Peter Stroh of Greater Detroit American Heritage River Initiative, was instrumental in galvanizing support on both sides of the border and in opening doors in both Canadian and U.S. governmental agencies.

Although the refuge was established in a relatively short period of time, it took much longer to achieve institutional legitimacy, particularly on the Canadian side. Building the refuge through cooperative agreements and voluntary initiatives was essential. The process of building the refuge was often slow due to institutional complexity and the nature of transactional costs (i.e. the time involved in coordinating efforts, the number of meetings, the amount of information required by stakeholders, etc.) in a major urban area. However, the consistent involvement of citizens and grassroots organizations in projects built local ownership and community support. This resultant broad-based citizen support for the refuge was essential to build capacity, recruit and retain stakeholder groups, secure necessary resources for projects, build public-private partnerships, achieve legitimacy, and establish local ownership. Of particular significance was the establishment of the Friends' Organization called the International Wildlife Refuge Alliance to build capacity, raise funds, increase outreach, and help form partnerships to advance conservation goals.

Both the Canadian Memorandum of Collaboration Agreement for the Western Lake Erie Watersheds Priority Natural Area and the U.S. Comprehensive Conservation Plan for the Detroit River International Wildlife Refuge called for enhanced cross border collaboration and cooperation, and for greater use of partnerships. As a result, both Canadian and U.S. governmental agencies worked together to achieve a compatible, mutually-shared, binational focus for fish and wildlife habitat protection, conservation, and rehabilitation in the shared waters of the Detroit River and western Lake Erie.

Sharing a compelling binational story was very important to generate support and enthusiasm in both the United States and Canada. This compelling story also had to be carried in the hearts and minds of all, and had to result in shared meaning. This story was that cooperative conservation is helping re-create gathering places for people and wildlife along the Detroit River and western Lake Erie. These unique conservation places are now a key factor in providing the quality of life demanded by competitive communities and businesses in the 21st Century. Equally important is that cooperative conservation is helping provide an exceptional outdoor recreational and conservation experience to nearly seven million people in the watershed. That, in turn, is helping develop the next generation of conservationists and sustainability entrepreneurs.

Literature Cited

Erg, B., Vasilijevic, M., McKinney, M. (Eds.). 2012. *Initiating effective transboundary conservation: A practitioner's guideline based on the experience from the Dinaric Arc.* Gland, Switzerland and Belgrade, Serbia.

Hartig, J.H., Robinson, R.S., Zarull, M.A., 2010. Designing a Sustainable Future through Creation of North America's only International Wildlife Refuge. Sustainability. 2(9):3110-3128. www.mdpi.com/2071-1050/2/9/3110/pdf

Hartig, J.H., Zarull, M.A., Ciborowski, J.J.H., Gannon, J.E., Wilke, E., Norwood, G., Vincent, A., 2007. *State of the Strait: Status and Trends of Key Indicators*. Great Lake Institute for Environmental Research, Occasional Publication No. 5, University of Windsor, Ontario.

Hartig, J.H., Zarull, M.A., Ciborowski, J.J.H., Gannon, J.E., Wilke, E., Norwood, G., Vincent, A., 2009. Long-term ecosystem monitoring and assessment of the Detroit River and Western Lake Erie. Environmental Monitoring and Assessment. 158, 87-104.

International Union for Conservation of Nature, World Commission on Protected Areas, 1997. Transboundary Protected Areas as a Vehicle for International Co-operation. Proceedings of the Parks for Peace Conference. Somerset West, South Africa.

Metropolitan Affairs Coalition, 2001. A Conservation Vision for the Lower Detroit River Ecosystem. Detroit, Michigan.

Norwood, G., Schneider, T., Jozwiak, J., Cook, A., Hartig, J.H., (Eds.), 2011. Establishing a quantitative target for common tern management in the Detroit River and western Lake Erie. Final report to Environmental Protection Agency. Grosse Ile, Michigan, USA.

Sandwith, T., Shine, C., Hamilton, L., Sheppard, D., 2001. *Tranboundary protected areas for peace and cooperation.* International Union for Conservation of Nature. Gland, Switerzland and Cambridge, UK.

Senge, P.M., 1990., *The Fifth Discipline: The Art and Practice of the Learning Organization.* Currency Doubleday Books, New York, New York.

Stein, J., Norwood, G., 2011. Detroit River Hawk Watch 2011 Season Summary. Detroit River International Wildlife Refuge, Grosse Ile, Michigan, USA.

U.S. Fish and Wildlife Service, 2005. *Comprehensive Conservation Plan and Environmental Assessment for the Detroit River International Wildlife Refuge.* Grosse Ile, Michigan, USA. http://www.fws.gov/midwest/planning/detroitriver/ (April 2013).

Vasilijevic, M., 2012. Transboundary conservation: An emerging concept in environmental governance. In: B. Erg, M. Vasilijevic, M. McKinney (Eds.), *Initiating effective transboundary conservation: A practioner's guideline based on the experience from the Dinaric Arc*, pp. 6-12. International Union for Conservation of Nature. Belgrade, Serbia.

Zbicz, D., 1999. Transboundary cooperation between internationally adjoining protected areas. In, D. Harmon (Ed.), *On the Frontiers of Conservation.* pp. 199-204. George Wright Society, Hancock, Michigan, USA.

CHAPTER 3

Roots of the Refuge: Standing on the Shoulders of Michigan United Conservation Clubs, Pointe Mouillee Waterfowl Festival, and the United Auto Workers

As noted in Chapter 2, the "Conservation Vision for the Lower Detroit River Ecosystem" (Metropolitan Affairs Coalition 2001) became the catalyst for the establishment of the Detroit River International Wildlife Refuge (DRIWR) in 2001. Although the refuge originated in 2001, it clearly has its roots in earlier efforts of conservation clubs, sportsmen's clubs, and industrial unions. This chapter will document how North America's only international wildlife refuge was built on the foundation established by the Michigan United Conservation Clubs (MUCC), Pointe Mouillee Waterfowl Festival, and the United Auto Workers (UAW) to protect an outdoor recreational heritage for all people, foster conservation, and control pollution.

Michigan United Conservation Clubs

After the Japanese attacked Pearl Harbor in 1941, the United States plunged into World War II. President Franklin D. Roosevelt led a "call to arm and support" the allied effort to win the war. In a historic speech, he referred to Detroit as the "Arsenal of Democracy" because of its rapid conversion from automobile to military manufacturing with the simple goal of helping win the war. In total, Detroit area companies received contracts worth about $14 billion or 10% of all U.S. military output in 1943. By 1944 Detroit was the leading supplier of military goods in the United States. Between 1942 and 1945 Detroit produced nearly $29 billion of military output. During these war years, over 600,000 people in the Detroit area were employed in this military production that ranged from ball bearings to bombers and tanks.

One can easily imagine how during these war years no one was thinking about protecting the environment. Literally, there were no pollution controls. Oil, heavy metals, and many other industrial contaminants were simply discharged into the Detroit River and the Rouge River, a major tributary that drains into the Detroit River, in substantial amounts. The U.S. Department of Health, Education, and

Welfare (1962) estimated that, during 1946-1948, 22.44 million liters (5.93 million gallons) of oil and other petroleum products were released into the Detroit and Rouge rivers each year.

Everyone now recognizes the potential harm from oil spills because of the 2010 BP oil spill in the Gulf of Mexico. How can we forget the pictures of oil flowing into the gulf and oil soaked carcasses of dead birds and other wildlife? Scientists have shown that about 3.8 liters (one gallon) of oil will contaminate approximately 3.8 million liters (one million gallons) of water (National Research Council, 1985; Central Contra Costa Sanitary District, 2008). By extrapolation, that means 22.44 million liters (5. 93 million gallons) of oil discharged annually during 1946-1948 was enough to pollute 22.44 cubic kilometers (5.93 trillion gallons) of water. The volume of U.S. and Canadian water in the western basin of Lake Erie is approximately 24.2 cubic kilometers (6.4 trillion gallons). That means that there was enough oil being discharged into the Detroit and Rouge rivers each year in 1946-1948 to pollute virtually all Michigan, Ohio, and Ontario waters in the western basin of Lake Erie.

This substantial oil pollution, primarily from industries lining the Rouge and Detroit rivers, took its toll in the winter of 1948. It was a particularly cold winter that year, allowing much of the Detroit River to freeze over (Hartig and Stafford, 2003). Only a few areas of open water remained. Hunt (1957) reported that prior to 1930 few waterfowl stayed in the Detroit River because ice generally covered the entire lower river. After 1930 portions of the lower river remained ice free because of the cumulative effect of warm water discharges (i.e. thermal pollution) from power plants, industries, and municipalities.

Ducks over-wintering on the Detroit River during 1948 headed for these open waters that were filled with oil and other petroleum products. The result was a massive winter duck kill due to the oil pollution, killing 11,000 ducks and geese. Under the orchestration of a well-known Downriver sportsman named Hy Dahlka and MUCC, local Downriver (i.e. a region of 23 communities in the lower Detroit River watershed) hunters collected the oil-soaked carcasses of waterfowl, threw them into their pickup trucks, drove them to the State Capitol in Lansing, dumped them on the Capitol sidewalk in protest, and held a press conference in opposition to the industrial water pollution of the Detroit River and the massive winter duck kill due to oil pollution (Figure 6). One of the leaders of the group claimed that the "ducks would serve to show State officials and the Legislature the need for an amendment to the present conservation law to make more severe pollution penalties." Photos were taken of the Michigan Governor tiptoeing through the piles of dead ducks and geese to get to work in the State Capitol (Fine, 2012). This single event has now been credited with starting the industrial pollution control program in Michigan (Cowles, 1975).

MUCC was created in 1937 by representatives of 35 sportsmen's clubs representing 6,000 club members (Dempsey, 2001). Its primary purpose was to speak in a "unified voice" on conservation issues in Michigan. By the end of the 20th Century it would speak for more than 100,000 members.

Figure 6. Oil-soaked carcasses of waterfowl from the lower Detroit River that were delivered to the State Capitol in protest of oil pollution in the Detroit River, 1948 (photo credit: State Archives of Michigan).

The catalyst for the creation of MUCC was a bill introduced in the Michigan legislature that would allow the Michigan Governor to appoint the Director of the State's conservation department. This action was seen as political and brought together representatives of the 35 sportsmen's clubs to protect Michigan's great outdoors. MUCC united sportsmen to collectively speak on behalf of conservation of natural resources and the hunting, fishing, and outdoor heritage that was dependent on their protection.

With increasing water pollution during and immediately following World War II, MUCC began calling for the cleanup of the Detroit River and other polluted water resources in Michigan. MUCC reported that "water pollution is considered the last important, uncontrolled, unregulated pagan practice continuing in the United States so far as natural resources are concerned" (Kistler, 1947). Hy Dahlka (Figure 7), an outspoken outdoorsman from Gibraltar, Michigan along the lower Detroit River who served on the MUCC Board of Directors for eight years, became Vice President, and served as President of MUCC in 1947, championed water pollution as "Michigan's public enemy No. 1" and noted that "the blight of pollution continues to transform some of our best streams into open sewers" (Dahlka, 1947). MUCC went on to identify the Detroit River as the most polluted river in Michigan at that time. Under considerable pressure from MUCC, Michigan Governor Sigler was forced to commission a special study of Michigan's water pollution in 1947 (Dempsey, 2001) that resulted in the Michigan legislature passing an "Anti-Pollution Law" in 1949.

Ken Dahlka (Figure 7), brother of Hy Dahlka, Chairman of the Pollution Committee of the Trenton Sportsmen Club that represented the Downriver communities, and active MUCC member, communicated the outrage and anger over pollution of the Detroit River:

> "I have lived in Trenton, Michigan, located 16 miles down the Detroit River at the head of Lake Erie for 36 years. I have watched the Detroit River change from one of the most beautiful rivers in the State of Michigan to the filthiest river in these United States, with the exception of the River Rouge. Pollution has increased to the extent that no longer are fish edible, paint on bottoms of boats

that use these waters is eaten off, aquatic vegetation has diminished to the extent that it has practically stopped the income of Muskrat trappers, of which hundreds of citizens participate annually. Also, swimming in these waters is a thing of the past."

Figure 7. Trenton Mayor Ken Dahkla and Gibraltar President and MUCC President Hy Dahlka shake hands at the border of their neighboring communities (photo credit: Gibraltar Historical Museum).

MUCC was relentless is its advocacy for controlling water pollution and protecting Michigan's outdoor heritage. Indeed, in a 1948 edition of its magazine called *Michigan Out-of-Doors*, MUCC (1948) called water pollution "an inexcusable waste of a most precious resource."

This outrage was further captured by Jack Van Coevering, conservation writer for the Detroit Free Press, who helped expand the anti-pollution crusade (Dempsey, 2001). The growing public awareness of water pollution and strong MUCC advocacy for water pollution control culminated in the 1948 press conference orchestrated by Hy Dahlka, MUCC, and Downriver duck hunters at the Michigan State Capitol that is now credited with starting the industrial pollution control program in Michigan (Cowles, 1975). Today, a Michigan Historical Marker stands at the Refuge Gateway in Trenton, Michigan commemorating the 1948 citizen activism that led to better industrial pollution controls and the recovery of the Detroit River. It should be noted that this historical marker at the Refuge Gateway was the first one established on the Michigan Conservation Trail by the Michigan Department of History, Arts, and Libraries. The cleanup of the Detroit River and the establishment of the DRIWR would probably not have happened without the early and strong anti-pollution advocacy of local sportsmen and the MUCC.

Pointe Mouillee Waterfowl Festival

If you fly in an airplane westward over Lake Erie, you come upon a massive, crescent-shaped, diked, wetland system called Pointe Mouillee (pronounced "Point Moo-yay"), sticking out like a "sore thumb" along the Michigan shoreline of the western basin. It got its name (i.e. meaning "wet point") from the French in 1700s. Much of this "wet point," that included many wetlands and islands, was formed by sediments transported from the Huron River. Native Americans and French trappers alike long marveled at this incredible delta and wetlands near the mouths of the Huron and Detroit rivers, particularly for their exceptional wildlife. French trappers were the first Europeans to come to the area seeking Beaver pelts during the Fur Trade Era. Long-standing residents have noted that, on a quiet day during certain times of the year, they could hear thousands of ducks and geese quacking and cackling from a kilometer away. Wild rice beds were so thick that men in waders had to remove them by hand so that hunters could pass in their boats (Marsh and Marsh, 1989). There were even times described by hunters when the flocks of ducks lifting off the wetlands were so big that they would change the color of the sky to black (Marsh and Marsh 1989). But Pointe Mouillee today bears little resemblance to the natural marshland that Native Americans, early French settlers, and early hunters marveled at – for it has gone through many, major, human-induced changes.

Today, 5.6 km (3.5 miles) of armored Pointe Mouillee dykes protect some of the most significant emergent and submergent marshland in Lake Erie, much like barrier islands protect coastal marshes along the shorelines of oceans (Figure 8A and B). This marshland contributes substantially to continentally-significant biodiversity in western Lake Erie, heralded in the North American Waterfowl Management Plan, the United Nations Convention on Biological Diversity, the Western Hemispheric Shorebird Reserve Network, and the Biodiversity Investment Area Program of Environment Canada and U.S. Environmental Protection Agency. It is literally an ecological jewel sandwiched between the major metropolitan areas of Toledo, Ohio and Detroit, Michigan.

Native Americans long recognized and hunted this area for the great bounty it provided their tribes. French explorers were the first Europeans to settle the region. Descendants of those French settlers lived in cabins on the islands and beaches, making a living by market hunting (i.e. commercial hunting of waterfowl and other game for money) and serving as guides for those who, in the middle of the 19th Century, began to come into this region to hunt (Fine, 2012).

Recognizing this exceptional wildlife resource, much like a jeweler recognizes a precious stone, William O. Hall purchased about 291 ha (720 acres) at $4.95 per hectare ($2 per acre) in 1872 for duck hunting. In 1875, William Hall partnered with E.H. Gilman to amass 485 ha (1,200 acres) of land at Pointe Mouillee and formed a shooting club that came to be known as the Big Eight Shooting Club. In 1879, this club was expanded to 10 wealthy businessmen from Michigan, New York, and Ohio, and renamed the Pointe Mouillee Shooting Club. As an exclusive shooting club, they employed punters (i.e. people who used flat-bottomed boats and a long pole to navigate in shallow waters) and cooks to facilitate meeting the wealthy hunters' recreational and culinary desires. It remained a private shooting club from 1875 to 1945. Each member was required to pay a fee (starting at $75 per year) for a share of the Club (Fine, 2012). Many of the

Club owners, over time, built nice cottages and cabins and the Club even had a clubhouse for dining and a gathering place called Rowdy Hall. The Club would acquire more land and expand over time, solidifying its reputation as one of the most prestigious hunting clubs in North America.

Figure 8A: Pointe Mouillee marshland before restoration in 1964 (photo credit: Southeast Michigan Council of Governments).

Figure 8B. Pointe Mouillee marshland after restoration in 2010 (photo credit: Southeast Michigan Council of Governments).

It remained a private hunting club until 1945 when the State of Michigan bought 1,055 ha (2,608 acres) from the Pointe Mouillee Shooting Club and established the Pointe Mouillee State Game Area. Over time, Pointe Mouillee State Game Area would expand and today includes 1,619 ha (4,040 acres) of marsh, shallow open water, dikes cropland, lake plain prairie, and lowland hardwood habitats (Cooley, 2011). Managed waterfowl hunting started in 1945 under the direction of the Michigan Department of Conservation (now called the Michigan Department of Natural Resources). Soon after, in 1947, avid hunters created the Michigan Duck Hunters' Tournament and the Pointe Mouillee Waterfowl Festival that became nationally recognized and continue today. This festival and tournament were started by Hy Dahlka, a well-known leader in the duck hunting community and President of Michigan United Conservation Clubs at the time, and four other duck hunters named Ed Lezotte, Bob Miller, Pete Petosky, and Lee Smits.

However, the quality of the hunting experience at Pointe Mouillee declined over time as high water levels in western Lake Erie eroded the natural protective barrier beaches by wind and wave action (U.S. Army Corps of Engineers, 2005). It was generally accepted that a combination of record high water levels in western Lake Eire and dams on the Huron River caused the erosion of large sections of the barrier beaches that led to loss of marsh habitat. Dams were built on the Huron River beginning in the 1930s, by Henry Ford and others, altering the flow and sedimentation patterns in western Lake Eire. In essence, less sediment was entering western Lake Erie, resulting in insufficient sediment accretion required to continually rebuild the barrier beaches that allowed the marsh vegetation to flourish. Finally, in 1952 a number of very large storms inundated and destroyed the marshland (Fine, 2012).

During the 1950s there was also an unsuccessful attempt to develop the land around Pointe Mouillee as a Port of Detroit (Fine, 2012). Sportsmen's Groups opposed this development on the grounds that Pointe Mouillee State Game was purchased with public money for the hunting and recreational benefit of all people.

Working-class sportsmen continued to advocate for preservation of Pointe Mouillee for public recreation. This advocacy culminated in 1960 when the Pointe Mouillee Waterfowlers' Association, affiliated with the Michigan United Conservation Clubs and the National Wildlife Federation, was created and dedicated to the preservation of the Lake Erie marshes (Fine, 2012).

There is evidence that as early as 1958 local sportsmen pressed for construction of a barrier dike to halt the degradation of the marsh by rising water levels. By the early-1960s local sportsmen were lobbying Secretary of Interior Stewart Udall, Michigan Governor G. Mennen Williams, and Congressman John Dingell (Figure 9). The first artificial dykes were constructed at Pointe Mouillee in 1963, reestablishing 147.7 ha (365 acres) of marshland in one unit (U.S. Army Corps of Engineers, 2005). Water levels within this marsh unit could now be managed independently of Lake Erie through the use of pumps and dewatering practices. It should be noted that managed waterfowl hunting had to be terminated temporarily in 1972 because much of the marsh had been destroyed by high water levels. This further outraged local hunters because of the loss of their sporting passion and they continued to advocate for the re-creation of a barrier dyke to be able to reestablish marshland to support waterfowl hunting. This advocacy for marsh restoration was undertaken by core group of duck hunters from the Pointe Mouillee Waterfowlers' Association, including Dick Micka, Hy Dahlka, Leonard Manaassa, Dick Whitwam, Ron Gorski, Jerry Ansman, and Don Girard. In 1974, Dick Micka was honored as "Conservationist of the Year" by MUCC for his advocacy and leadership on marsh preservation, including making presentations before the Michigan Natural Resources Commission and the U.S. Army Corps of Engineers. The original plan to construct barrier dykes that would allow the marsh to once again flourish was put forward by Clyde Odin of U.S. Fish and Wildlife Service and Pete Petosky of Michigan Department of Natural Resources.

As a result of public outrage over open water disposal of contaminated dredged sediment in the Great Lakes, President Lyndon Johnson signed an Executive Order banning such practices in 1966 (Buffalo Niagara Riverkeeper, 2011). Governments had no choice but to seek places to confine contaminated sediment. The solution was the construction of a confined disposal facilities or CDFs for confining contaminated sediment. The unique confluence of strong hunter interest in restoring the marshes of Pointe Mouillee and governmental need to confine contaminated sediments led to one of the largest marsh restoration projects in the world. The solution was to construct 2.8 km^2 (700 acres) of dyked disposal areas consisting of five cells that would eventually contain 13.8 million m^3 (18 million cubic yards) of contaminated dredged material from the Rouge and Detroit rivers in southeast Michigan (U.S. Army Corps of Engineers, 2005). With the support of U.S. Senator Robert Griffin, the Pointe Mouillee CDF was authorized by the River and Harbor Act of 1970 (Public Law 91-611, Section 123) and constructed by the U.S. Army Corps of Engineers. The 5.6 km (3.5 miles) of dyked disposal cells were constructed during 1976-1981 at a cost of $45.5 million (U.S. Army Corps of Engineers, 2005). These dyked disposal cells would also double as a wave barrier for reestablishing marshland managed by the Michigan Department of Natural Resources as part of Pointe Mouillee State Game Area in support of re-creating the world-class hunting heritage. It was an innovative "win-win" situation for addressing an environmental problem and restoring marshland for waterfowl hunting.

Figure 9. Hy Dalhka, U.S. Secretary of Interior Stewart Udahl, Ken Dalhka, Michigan Governor G. Mennen Williams, and Congressman John D. Dingell (left to right) discuss the need to restore barrier dikes at Pointe Mouillee, early-1960s (photo credit: Pointe Mouillee WaterfowlFestival).

The perimeter and cross dikes were constructed of clay and various sized limestone, along with filter fabric, to prevent leakage of contaminated material and to resist wave action.

Again, this project was recognized at the time as one of the world's largest marsh restoration efforts. Conservation and outdoor recreational benefits of this Pointe Mouillee marsh restoration project included (Pointe Mouillee Waterfowl Festival, 2012):

- Increasing annual production of waterfowl (peak annual fall waterfowl migration estimated at 20,000-26,000 birds);
- Sustaining significant hunting opportunities (waterfowl hunting estimated at 6,000 user trips annually; total hunting effort estimated at 10,000 user trips annually);
- Supporting trapping (13,710 Muskrats trapped in 2006);
- Sustaining exceptional biodiversity (e.g. Black-bellied Plover – *Pluvialiss quatarola*; Glossy Ibis – *Plegadis falcinellus*; Whimbrels – *Numenius phaeopus*; Bald Eagles, Osprey, Eastern Fox Snake – *Elaphe gloydi*);
- Providing outstanding fishing opportunities; and
- Offering exceptional birding opportunities recognized by the International Shorebird Convention (e.g. 10,000-20,000 shorebirds using the area; Osprey, Bald Eagles, and Black Necked Stilts (*Himantopus mexicanus)* nesting; one of the largest Black-crowned Night Heron (*Nycticor axnycticorax*) rookeries in the Midwest with 225 reported in 2011).

Today, Pointe Mouillee State Game Area continues to be managed by the Michigan Department of Natural Resources to: restore and maintain biological communities; and provide public use opportunities through practices and improvements that do not disturb existing unique features and which complement natural processes and local area ecology. It is also managed cooperatively with U.S. Fish and Wildlife Service through a Memorandum of Understanding for the DRIWR.

Pointe Mouillee State Game Area provides close-to-home, world-class hunting opportunities to nearly seven million people in a 45-minute drive. Indeed, this area was designated by Ducks Unlimited in 2011 as one of the top ten urban waterfowl hunting areas in the United States. Interestingly, it is also well known locally for Muskrat trapping because Catholics in the Downriver area have had a long-standing tradition of eating Muskrat under a traditional exemption to the eating of meat during Lent. Pointe Mouillee also provides world-class wildlife observation opportunities for waterfowl, shorebirds, and other wetland wildlife, and exceptional hiking and biking for outdoor enthusiasts.

Michigan Department of Natural Resources (Cooley, 2011) estimated that in the Pointe Mouillee State Game Area during 2011 there were:

- 6,000 user trips (i.e. the total number of trips taken by all individuals over an increment of time like one year) for waterfowl hunting;
- 300 user trips for deer hunting;
- 3,000 user trips for small game hunting;
- 30,000 user trips for fishing;
- 6,000 user trips for wildlife viewing;
- 8,000 user trips for the Pointe Mouillee Waterfowl Festival;
- 500 user trips for educational tours; and
- 3,000 user trips for other activities like boating.

That means that during 2011 it was estimated that there were 56,800 user trips for outdoor recreational and educational activities. Waterfowl hunting remains outstanding and continues to be celebrated through the annual Michigan Duck Hunters' Tournament. In 2011 and 2012, considerable management efforts were expended on controlling the invasive reed-like plant *Phragmites* through use of selective herbicides and prescribed fires, and on wetland restoration and dyke rehabilitation with $500,000 of Great Lakes Restoration Initiative funding.

The annual tradition of the Pointe Mouillee Waterfowl Festival has also been sustained since it was first established in 1947. Although contestants were scarce in the first tournaments, participation steadily grew and peaked in the 1960s-1970s at 15,000-20,000 (Figure 10). Today, it still attracts 8,000-10,000 people. Over the years it has been described as one of the most unusual tournaments in the United States because it has focused on testing duck hunters' abilities, including punt boat racing, decoy making, layout shooting, dog training, duck calling, decoy head whittling, and duck feather picking. Sponsored by the Pointe Mouillee Waterfowl Festival Committee on the first weekend after Labor Day and right before duck hunting season opens, the festival has become an annual tradition with all event proceeds going to the restoration and maintenance of Pointe Mouillee State Game Area.

Figure 10. Pointe Mouillee Waterfowl Festival in Brownstown, Michigan, 1977 (photo credit: Pointe Mouillee Waterfowl Festival Committee)

Probably one of the most important outcomes of the annual Pointe Mouillee Waterfowl Festival and its associated Michigan Duck Hunters' Tournament, however, is that the hunters have not only passed along their hunting passion and tradition, but also laid the foundation for the establishment of the DRIWR. Indeed, without the strong hunting and conservation heritage passed down through many generations by the Pointe Mouillee Waterfowl Festival and its Michigan Duck Hunters' Tournament, we may not even have had an international wildlife refuge.

United Auto Workers (UAW)

It was the early-1960s and industry was king. Detroit was the Motor City and manufacturing and industrial jobs were the top priority. It was also before Earth Day, the U.S. Clean Water Act, the U.S. Endangered Species Act, and the U.S.-Canada Great Lakes Water

Quality Agreement. The Detroit River was still viewed as a working river and pollution was considered part of the cost of doing business (Hartig, 2010). The wastewater treatment plants had just primary treatment and industrial pollutants were still being discharged in substantial amounts. Oils slicks were still a common sight. Indeed, in 1960, another 12,000 ducks and geese died on the river from oil pollution (Hartig and Stafford, 2003). Bald Eagles, Peregrine Falcons, Osprey, Lake Sturgeon, and Lake Whitefish were no longer found in and along the Detroit River, and Walleye were considered in a crisis state (Hartig et al., 2009). Picture a river that ran grey to brown to black, and that was considered a cesspool and health hazard by governments.

Again, most people were disconnected from the Detroit River, unaware of the severity and extent of water pollution, or apathetic about its grossly polluted state. Few people recognized that we were on the precipice of an environmental "tipping point." There were very few organized groups campaigning against water and air pollution in that area until the UAW became an advocate for recreation, conservation, and pollution control.

In Michigan, automakers were long involved in sportsmen's associations, dating back to the 1920s and 1930s (Montrie, 2008). Starting in the 1920s, Montrie (2008) argues that a growing number of male autoworkers and other factory workers living in and around Detroit turned to fishing and hunting to counteract their sense of estrangement from the natural world. To support those activities they joined sportsmen's clubs. Automobile and other factory workers' first-hand experiences with water and air pollution put them in the forefront of efforts to control pollution and benefit from outdoor recreational pursuits like fishing and hunting (Montie, 2008).

The UAW was chartered in 1935 in Detroit under the auspices of the American Federation of Labor. Its primary concern was wages, hours, and pensions. As the UAW grew in size and influence through the 1940s, worker recreation became more important. When Walter Reuther won control of the UAW in 1946, he made the UAW a major force in the automotive industry and Democratic Party. He was an old-style socialist in his early life and drew on this ideological background to advance and evolve a liberal agenda that embraced many social issues of the day, including worker education and recreation (Montrie, 2008).

Reuther loved the water and had taken his family to a cabin on Higgins Lake in Michigan's northern Lower Peninsula each summer. When he later moved to a 2.02 ha (five-acre) plot on Paint Creek, a tributary of the Clinton River in Macomb County, Michigan, he helped organize Boy Scout cleanups of the creek and advocated for money to build wastewater treatment plants (Dempsey, 2001). His daughter remembered him telling her as a 12-year old child how humans were destroying nature's balance and she later, as an adult, felt that his "concern for the earth was the ripened fruit of his humanity" (Dempsey, 2001).

Reuther's long-standing commitment to worker recreation and the environment helped establish the foundation for developing autoworkers' environmental consciousness that led the UAW to become one of labor's foremost advocates for pollution controls (Montrie, 2008). Indeed, it was the UAW's interest in promoting recreational and leisure activities for its workers, and the workers' growing interest in protecting the environment, that led to the development of a working-class environmental consciousness and a labor environmentalism which we all benefit from today (Montrie, 2008).

This working-class environmental consciousness and labor environmentalism were manifested in a statement adopted in 1967 by the UAW's International Executive Board:

> "As we bargain for more time off the job in the form of more paid holidays, longer vacations and early retirement, we must also be concerned that this time can be utilized in the kinds of surroundings that will bring some measure of pleasure, satisfaction and enjoyment to our members and their families. Our Union has demonstrated time and time again that its responsibility to its members does not end with the punching out at the time clock, and that the conservation issue is as vital to us as any we have previously undertaken."

The International Executive Board then set up a special committee, headed by President Walter Reuther, to work with staff in the areas of conservation and recreation, and to implement a pilot program for anti-pollution activities in the Downriver area. Walter Reuther turned to a passionate individual who would be a strong advocate on issues pertaining to conservation and recreation.

She was the ninth of 13 children born to Slovakian immigrants in 1915. At an early age, she inherited a strong work ethic from her father and a deep appreciation for the beauty of the environment from early experiences on camping and fishing trips (Dekutoski, 2008). Her family first resided in Sykesville, Pennsylvania where her father owned a butcher shop. When the butcher shop failed, the family moved to Detroit, Michigan in search of better work opportunities.

She attended Northeastern High School in Detroit and became a good athlete, participating in softball, basketball, and soccer. However, sports were not her only passion. She was an honor roll student, and was active in the Loyalty Dramatic Club, the Mermaids Club, and Girls' Athletic Association. She also found time to be the sports editor for her high school newspaper.

Her name was Olga Madar (Figure 11) and she would go on to become a prominent leader of the UAW and a champion of neglected elements of her world, including the environment, recreational opportunities for urban areas, equity for all, and caring for senior citizens (Dekutoski, 2008). After graduation from high school in 1933, Madar went to Michigan Normal College (now Eastern Michigan University) to become a teacher. While going to college she needed to support herself and worked in auto plants from 1933 to 1935. In an oral history conducted with Roth (1994), Olga Madar noted that the only reason she got a job in the auto plants was because she could play softball. Through those auto plant jobs she became exposed to the assembly line and the kind of conditions that were common before unions.

After graduation from Michigan Normal College, she went on to teach in the Flat Rock school system in southeast Michigan. She taught geography, history, civics, and physical education for nearly three years. Her pay was relatively low and did not increase over three years. She soon realized that she could work at Ford Motor Company for higher wages. She went back to work at the auto plant and also pursued her interest in recreational programs. Madar joined the UAW in 1944 and shortly after was appointed Director of Recreation for UAW Local 50 (Willow Run, Michigan). Madar was deeply affected by the poor conditions many laborers were experiencing (Dekutoski, 2008). She organized softball teams and bowling leagues. Her leadership in recreational programs caught the attention of UAW President Walter Reuther, who worked with her to eliminate racial discrimination in organized bowling leagues and to ensure equal rights for women (Dekutoski, 2008). She went on to become the Director of the Recreation Department of the UAW in 1947. She further demonstrated her passion for recreation by serving

Figure 11. UAW Vice President Olga Madar, late-1960s (photo credit: Walter P. Reuther Library, Wayne State University).

an eight-year term as a Detroit Parks and Recreation Commissioner beginning in 1959.

Olga Madar became the first woman elected to the International Executive Board of the UAW in 1966 and would go on to serve two terms as Vice-President of the UAW (Figure 11). For Madar, conservation and preservation of open space and natural resources were clearly tied to recreational activities. The UAW International Executive Board created the Department of Conservation and Resource Development in 1967, under the direction of Madar, to work with advocacy groups in the United States and Canada in lobbying legislators and developing educational programs to protect the environment and conserve natural resources. For example, she provided key leadership to the UAW in national efforts to ban DDT and to provide more advanced treatment at wastewater treatment plants polluting the Great Lakes (Dekutoski, 2008). By 1967 she had formed a coalition of labor, conservation clubs, and garden clubs to ensure enforcement of water and air pollution laws. The UAW, under the leadership of Olga Madar and Walter Reuther, was front and center in the efforts to clean up Lake Erie. In a 1968 letter to then Secretary of Interior Stewart Udall (Dekutoski, 2008), Olga Madar stated, "We must overcome the [pollution] problem if we are to preserve a living environment worthy of and advantageous to the citizens of a free society."

In 1969, Madar worked through the UAW's Conservation and Resource Development Department to create the Downriver Anti-Pollution League. Its goals were to:

- Strengthen anti-pollution legislation and standards;
- Strengthen enforcement efforts;
- Help citizens work with organizations and groups to make their voices heard;
- Make governmental and other organizations more responsible to the public; and
- Obtain equitable distribution of costs and consequences of pollution prevention and cleanup.

Swan (1992) notes that during that time, there were no organized groups campaigning against water and air pollution in that area until the UAW created the Downriver Anti-Pollution League in the late-1960s.

Two part-time University of Michigan interns were hired that year to work with local UAW staff and members to organize the communities (i.e. Ecorse, Lincoln Park, River Rouge, and Wyandotte) through the Downriver Anti-Pollution League to fight pollution. Hundreds of UAW members and their families attended the first meeting. One of their early environmental educational efforts was sponsoring a tour of sources of pollution in the area to raise public awareness. Another effective initiative organized by the Downriver Anti-Pollution League was a U.S.-Canadian wake mourning the death of the Detroit River and Lake Erie on the first Earth Day (April 22, 1970). Boats from both countries met in the middle of the river to hold the wake and place a floral wreath on the murky waters of the Detroit River (Figure 12). Mrs. Peter Naccrato, Vice-Chairperson of the Downriver Anti-Pollution League, quickly retrieved the wreath from the Detroit River saying "I don't want the flowers to get contaminated." An April 22, 1970 Earth Day Resolution issued by the Downriver Anti-Pollution League noted:

> "We, the citizens of Michigan, especially those of us living along the Detroit River, realize that the pollution from American companies along the river affects not only us but our Sisters and Brothers on the Canadian side. We are demanding that our government take more stringent measures against corporations, municipalities, and other sources of pollution."

The event received much media coverage.

The UAW continued to organize its members around a campaign to increase environmental awareness and action on the part of industrial workers and their families. They adeptly used union organizing skills and non-formal educational methods to accomplish their goal. The compelling rationale given for this environmental campaign is presented below:

> "American workers, perhaps more than the rest of the nation, have good reason to be foes of pollution. They have confronted it, resisted it, and to a dangerous degree have had to endure it over decades on the job. In-plant hazards have increased with the proliferation of new toxic substances in recent years. Moreover, workers and their families are most apt to be exposed to pollution released by industry in the surrounding community, for they are less likely than executives and professional workers to live in residential suburbs..... They are not advocates of pollution, yet their economic circumstances require them to think first of jobs, paychecks and bread on the table..... There must be action to assure American workers and their families of a valid alternative to paychecks earned through working and living in a polluted environment. That alternative, put simply, is the alternative of jobs, paychecks, bread on the table – and a clean environment."

Figure 12. Downriver Anti-Pollution League holds wake on the Detroit River on the first Earth Day, April 22, 1970 (photo credit: Walter P. Reuther Library, Wayne State University).

Perhaps the following statement from a 1970 resolution, adopted at the 22nd UAW Constitutional Convention in Atlantic City, New Jersey, demonstrates best the UAW's role in raising environmental awareness and consciousness:

> "For too long our nations have measured affluence by the quantity of goods produced without regard to the damage done to the quality of man's environment. In the pell-mell rush for technological advancement, we have paid too little attention to the destruction wreaked upon the natural world about us. We must learn to master the forces of technological changes and bend them in constructive, creative and responsible directions so that we can continue to increase our material affluence without corrupting or destroying our living environment."

By the end of the 1970s, the labor environmentalism that autoworkers had worked so hard to build had waned (Montire, 2008). However, the seeds of labor environmentalism that were sown in the 1960s clearly took hold in our annual celebration of Earth Day, the U.S. Clean Water Act, the U.S. Endangered Species Act, the U.S.-Canada Great Lakes Water Quality Agreement, and now the DRIWR. Indeed, Gaylord Nelson, the founder of Earth Day, noted that the UAW and other labor unions became important constituencies in the modern environmental movement (Nelson, 2010).

Concluding Thoughts

MUCC, Pointe Mouillee Waterfowl Festival, and the UAW each played critical roles in furthering conservation, environmental protection, and outdoor recreation in this ecological corridor. There were also times when these three entities collaborated to further the cleanup and conservation of natural resources in the Detroit River and western Lake Erie. Not only were these three entities advocates for conservation, environmental protection, and outdoor recreation for all, but they were a voice for future generations. Clearly, their legacy can be seen today in a cleaner Detroit River and western Lake Erie, the recovery of sentinel fish and wildlife species, the conservation of continentally-significant natural resources like Pointe Mouillee State Game Area and the DRIWR, and world-class fishing, hunting, birding, boating, paddling, and other outdoor recreational opportunities for nearly seven million people in a 45-minute drive, and for many annual visitors.

Human knowledge, understanding, and development are most often viewed as cumulative affairs. Indeed, Isaac Newton recognized this in his classic quote:

"If I have seen further, it is by standing on the shoulders of giants."

Therefore, we need to recognize that the DRIWR and the many benefits derived from it are the result of "standing on the shoulders of giants" like MUCC, Pointe Mouillee Waterfowl Festival, the UAW, and others. Indeed, we must recognize that we are today leveraging the historical conservation and environmental protection work of these "giants" in building North America's only international wildlife refuge, and that we couldn't have done this on our own. And for this we pay homage to these organizations and their members, and we are most grateful!

Literature Cited

Buffalo Niagara Riverkeeper, 2011. 2008-2010 Triennial Report. Buffalo, New York.

Cooley, Z., 2011. Pointe Mouillee State Game Area Annual Report. Michigan Department of Natural Resources, Rockwood, Michigan.

Cowles, G., 1975. Return of the river. Michigan Natural Resources. 44(1), 2-6.

Dahlka, H., 1947. Message from the President.Michigan-Out-of-Doors. 1(1), 13.

Dekutoski, D.A., 2008. Olga Madar: Lifelong Activist. M.A. Thesis, Wayne State University, Detroit, Michigan, USA.

Fine, L., 2012.Workers and the Land in U.S. History: Pointe Mouillée and Detroit Downriver Detroit Working Class in the Twentieth Century. Labor History 53 (3), 409-434.

Hartig, J.H., 2010. *Burning Rivers: Revival of Four Urban-Industrial Rivers That Caught on Fire.* Ecovision World Monograph Series, Aquatic Ecosystem Health and Management Society, Burlington, Ontario, Canada and Multi-Science Publishing Company, Ltd., Essex, United Kingdom.

Hartig, J.H., Stafford, T., 2003. The Public Outcry over Oil Pollution of the Detroit River. In, J.H. Hartig (Ed.). *Honoring Our Detroit River, Caring for Our Home.* pp. 69-78. Cranbrook Institute of Science, Bloomfield Hills, Michigan, USA.

Hartig, J.H., Zarull, M.A., Ciborowski, J.J.H., Gannon, J.E., Wilke, E., Norwood, G., Vincent, A., 2009. Long-term ecosystem monitoring and assessment of the Detroit River and Western Lake Erie. Environmental Monitoring and Assessment. 158, 87-104.

Hunt, G., 1957. Causes of mortality among ducks wintering on the lower Detroit River.Ph.D. Dissertation, University of Michigan, Ann Arbor, Michigan, USA.

Kistler, G.E., 1947. Michigan's public enemy No. 1 – Pollution.Michigan-Out-of-Doors.1(1), 2 and 23.

Marsh, J. Marsh, B., 1989. The old Pointe Mouillee Shooting Club. Michigan Out-of-Doors. 43 (December), 18-21.

Metropolitan Affairs Coalition, 2001. A conservation vision for the lower Detroit River ecosystem. Detroit, Michigan, USA.

Michigan United Conservation Clubs, 1948. Michigan's shame – pollution. Michigan-Out-of-Doors: 2(1), 3 and 20.

Montrie, C., 2008. *Making a Living: Work and the Environment in the United States.*The University of North Carolina Press, Chapel Hill, North Carolina, USA.

Nelson, G., 2010. *The UAW Steps up for Earth Day.* Nelson Institute for Environmental Studies, University of Wisconsin, Madison, Wisconsin, USA. http://www.nelsonearthday.net/collection/coalition-uawflyer.htm (April 2013).

Roth, S. 1994. Oral history interview with Olga Madar, May 21, 1994.Walter P. Reuther Library, Wayne State University, Detroit, Michigan, USA.

Swan, J.A., 1992. *Nature as a Teacher and Healer: How to Reawaken your Connection to Nature.* Villard-Random House, New York, New York, USA. www.jamesswan.com/natureasteacherandhealer.html (April 2013).

U.S. Army Corps of Engineers, 2005. Pointe Mouillee Confined Disposal Facility and State Game Area. Detroit, Michigan, USA.

U.S. Department of Health, Education, and Welfare, 1962.Pollution of the navigable water of the Detroit River, Lake Erie and Their Tributaries within the State of Michigan. Detroit, Michigan, USA.

CHAPTER 4

Public-Private Partnerships for Conservation

A partnership is generally understood to be an arrangement where parties agree to cooperate to advance their mutual interests. Sometimes a partnership is established because the desired task cannot be accomplished alone or to avoid duplication of efforts; other times a partnership is formed to work more effectively, more efficiently, and more economically. For those who have worked in very challenging partnerships, they will probably find humor in the often cited quote by Quentin Crisp:

> *It is explained that all relationships require a little give and take. This is untrue. Any partnership demands that we give and give and give and at the last, as we flop into our graves exhausted, we are told that we didn't give enough.*

In the urban conservation field, partnerships are critical because of the number of stakeholders involved and impacted, and the complexity of necessary solutions. Most urban areas, by their very nature, have large human population densities, substantial housing and commercial developments, usually considerable urban sprawl, and frequently much industrial development. This urbanization and industrialization have also frequently left a legacy of lost wildlife, wooded areas, and wetlands, and often a legacy of contaminated environments upon which our cities and their ecosystem health depend. To address these challenges and indeed bring conservation to urban areas will require partnerships of many kinds and at many levels. No one can do it alone. Today, the conservation challenges in urban areas are far too big for any single agency, organization, or interest group to surmount – they require all to do their part and to work together.

Legitimacy

The use of partnerships for urban conservation can be opportunistic or random, or can be strategic. In the case of the Detroit River

International Wildlife Refuge (DRIWR), the use of public-private partnerships for conservation purposes was strategic and given legitimacy through enabling legislation, formal plans, and collaborative agreements.

Legitimacy is generally understood to mean justified and genuine. When considering the use of public-private partnerships to grow and build an international wildlife refuge, it is worthwhile to think about it from the perspective of legitimacy. From an institutional perspective, MacKenzie (1996) notes that legitimacy creates spaces in which to bargain, negotiate, and persuade other organizations and agencies to implement a program (e.g. like an urban wildlife refuge) where sanctions and incentives are otherwise absent. Perceptions of legitimacy are important when many organizations with overlapping authority work together, even more so when the initiative is new and may or may not have a lot of support. In the case of the DRIWR, the first measure of legitimacy came when the Conservation Vision was signed in 2001 on behalf of Canada by then Canadian Deputy Prime Minister Herb Gray and behalf of the United States by U.S. Congressman John Dingell. Next, the DRIWR received another measure of legitimacy when it was created by U.S. federal act in 2001 and when Canada agreed to use a number of existing Canadian Acts to support the concept of the international wildlife refuge on the Canadian side. The early and consistent political support for the refuge cannot be underestimated. Having Canadian Members of Parliament like Herb Gray, Susan Whelan, and Jeff Watson, and U.S. Congressmen and U.S. Senators like John Dingell, John Conyers, Carl Levin, and Debbie Stabenow, give strong political support contributed to the perception of legitimacy. Other factors that contributed to legitimacy included: strong public participation and involvement as called for in the Comprehensive Conservation Plan (U.S. Fish and Wildlife Service, 2005); and early funding to support projects like State of the Strait Conferences, shoreline restoration, a bird driving tour map, land acquisition, etc.

This perception of legitimacy was instrumental in gaining the support of corporations, other businesses, communities, nonprofit organizations, and others to experiment with public-private partnerships for conservation, particularly on the U.S. side. In essence, this helped establish trust among stakeholders, helped foster good working relationships, and helped build a record of success.

Clearly, in order to achieve the conservation goals of the refuge, partnerships had to be established at all levels of government, with the private and nonprofit sectors, with academic institutions, and other key stakeholder groups. The Detroit River International Wildlife Refuge Establishment Act of 2001 states that one of the three key purposes for which the Refuge was established is to facilitate partnerships among the U.S. Fish and Wildlife Service, Canadian national and provincial authorities, State and local governments, local communities in the United States and in Canada, conservation organizations, and other non-Federal entities to promote public awareness of the resources of the Detroit River. The Establishment Act also states that the U.S. Secretary of Interior is authorized to enter into cooperative agreements with the State of Michigan, or any political subdivision thereof, and with any other person or entity for the management, in a manner consistent with this Act, of lands that are owned by such State, subdivision, or other person or entity and located within the boundaries of the refuge. Further, the U.S. Secretary of Interior is directed to promote public awareness of the resources of the DRIWR and encourage public participation in the conservation of those resources.

Even the Comprehensive Conservation Plan for the refuge recognizes that "partnerships will be a key element for the future of the Refuge" (U.S. Fish and Wildlife Service, 2005). This plan, that guides the management of the refuge for 15 years, states that the Service will actively seek to develop partnerships with public and private organizations in an effort to support conservation, environmental education, and habitat restoration initiatives within the authorized boundary of the refuge. Indeed, the "Number 1" goal identified in the Comprehensive Conservation Plan states that the Service will establish partnerships involving communities, industries, governments, citizens, nonprofit organizations, and others to manage and promote the refuge consistent with the Plan's vision statement and the Establishment Act of 2001 (U.S. Fish and Wildlife Service, 2005).

Further, the Canadian Memorandum of Collaboration Agreement for the Western Lake Erie Watersheds Priority Natural Area states that one of the primary objectives is to "improve the effectiveness and efficiency of natural resource management in the area of interest through enhanced collaboration and cooperation among the Parties and other federal departments, provincial ministries, and municipalities, including any of their relevant organizations."

These laws, formal plans, and collaboration agreements gave legitimacy to establishing public-private partnerships for conservation, but also helped create the framework and necessary conditions for conservation partnerships. Presented below are some selected examples of public-private partnerships used in conservation work within the DRIWR, including compelling results achieved.

Planning for Conservation of Lands and Biodiversity

As noted In Chapter 2, extensive U.S.-Canada partnerships were used to develop the 2001 Conservation Vision and the Comprehensive Conservation Plan for the DRIWR. However, there are other examples of the use of effective partnerships for planning for conservation of lands and biodiversity in the corridor. Table 6 presents three examples of other planning initiatives that involved extensive partnerships. In the case of the Lake Erie Biodiversity Strategy (Pearsall et al., 2012) and the Essex Forests and Wetlands Conservation Action Plan, partners from both Canada and the U.S. were actively involved in the planning efforts. These planning initiatives involved convening regular meetings, sharing information, coordinating program activities, setting priorities, and, in some cases, undertaking shared projects from an ecosystem perspective. Such transboundary planning efforts have helped lay the foundation the DRIWR. In the case of the Essex Region Biodiversity Conservation Strategy (Essex Region Conservation Authority, 2002), there was limited U.S. involvement due to the exclusive focus on Essex County region of Ontario. In the future, as institutional arrangements mature and additional conservation successes are achieved, there will undoubtedly be more effort expended on transboundary conservation planning. Clearly, much remains to be done institutionally through partnerships to fully realize the transboundary conservation and ecosystem-based management goals of the two countries.

Land Donations

Mud Island is one of over 30 islands in the Detroit River and is located within the City of Ecorse, Michigan. It was recorded on navigational maps as far back as 1796. In the 1800s, it was owned by the U.S. Government. During this time, strong river currents and winter ice scour eroded Mud Island to a narrow strip of land about 800 meters (874 yards) long and about 100 meters (109 yards) wide. By 1912, the island had diminished in size to less than 0.2 ha (0.5 acres).

In 1924, a fisherman named Nicholas Hohfelt purchased it and built a house on it. But fishing wasn't all that Mr. Hohfelt was interested in. Adjacent to his house was a partially submerged boathouse that was a terminus for an underwater cable system that carried illegal alcohol along the river bottom from Canada during Prohibition (Nolan, 1999). During the Prohibition Era, gangs grew increasingly violent and brazen, and gunfire between federal agents and gangsters made it a dangerous place to live. Finally, Hohfelt had to abandon the island in 1930 due to the problems with the rum-runners and smugglers.

In 1945, National Steel Corporation purchased Mud Island and its submerged shoals from Hohfelt. In 1962, the U.S. Army Corps of Engineers expanded the island to approximately 8 ha (21 acres) using clean dredged materials from the shipping channel of the Detroit River. In the 50 years since the placement of clean dredged material on Mud Island, it has become completely forested with maples, cottonwood, and ash trees, and has become a haven for wildlife. The island now serves as important stopover habitat for birds and the shoals support extensive beds of Wild Celery (*Vallisneria americana*) that serve as an important staging area for waterfowl and an important spawning and nursery ground for fishes.

In 2001, National Steel Corporation donated Mud Island to the U.S. Fish and Wildlife Service for the refuge. No industrial operations were ever conducted on the island. Today, it stands approximately 61 m (200 feet) from John Dingell Park in Ecorse and is seen by tens of thousands of annual park visitors. Mud Island was literally a gift from National Steel Corporation to the U.S. Fish and Wildlife Service for the refuge and is now protected in perpetuity for migratory birds, fish, and other wildlife. This early donation of Mud Island was an excellent example of how a government agency could work with a private corporation to benefit wildlife and future generations of anglers and hunters who use its shoals. It literally provided an example of public-private partnerships for conservation for other corporations.

At the lower end of the Detroit River in the City of Gibraltar lies a 144.9-ha (358-acre), crescent shaped wetland complex surrounding Gibraltar Carlson High School and Shumate Middle School. It is an incredible "nursery of life" that provides habitat for thousands of

Table 6. Examples of biodiversity and land conservation planning initiatives for the Detroit River and western Lake Erie corridor.

Plan/ Strategy	Focus	Partners
Essex Region Biodiversity Conservation Strategy	This strategy produced a spatial database of all natural areas in the Essex region and conducted an analysis of the terrestrial, wetland, and riparian habitats to identify the extent of existing natural vegetation. This was then used to prioritize opportunities for habitat rehabilitation and enhancement. Strategic planning for the rehabilitation and restoration of ecosystem features focuses on identifying high priority opportunities to help restore or improve environmental features and ecological functions that have been lost or degraded. The objective of these measures is to increase the size, extent, and quality of key natural heritage features, natural corridors, and greenway linkages, thereby improving the ecosystem diversity and ecological functions of the watersheds.	Essex Region Conservation Authority, Environment Canada, Essex County Woodlot Owners Association, Ducks Unlimited Canada, Ontario Ministry of Natural Resources, Essex County Stewardship Network, Essex County Federation of Agriculture, County of Essex, Essex County Field Naturalists, University of Windsor, Canadian Wildlife Service, Little River Enhancement Group, Parks Canada, Carolinian Canada, Citizen Environment Alliance, Sternsman International, and Prince, Silani & Associates
Essex Forests and Wetlands Conservation Action Plan	This action plan calls for a coordinated approach to conservation and stewardship of critical forest and wetland habitats in Essex County. Action plan goals include: maintain existing and establish new functional ecological linkages between core conservation areas; complete securement of core conservation areas; maintain and recover viable populations of Species at Risk, including reducing anthropogenic mortality of all reptile species; manage invasive species populations so no net increase in population density occurs; increase natural cover through restoration to a total of 12% of the landscape; enhance community support and understanding of Essex forests and wetlands and to promote community participation in its conservation, including enforcement of policies and regulations; enhance information and monitoring of biodiversity values, natural processes and threats; and support and enhance conservation partnerships across the Natural Area.	Nature Conservancy of Canada, Essex Region Conservation Authority, Essex County Stewardship Network, Parks Canada, Bird Studies Canada, Canada South Land Trust, Carolinian Canada Coalition, Tallgrass Ontario, Essex County Stewardship Network, Ontario Ministry of Natural Resources, Parks Canada, and the Michigan Chapter of The Nature Conservancy

Table 6 Con't

Lake Erie Biodiversity Conservation Strategy (LEBCS)	LEBCS is a U.S.-Canada initiative designed to identify specific strategies and actions to protect and conserve the native biodiversity of Lake Erie. The goals of this planning process included: assemble available biodiversity information for Lake Erie; define a binational vision of biodiversity conservation for Lake Erie; develop shared strategies for protecting and restoring critical biodiversity areas; describe the ways in which conservation strategies can benefit people by protecting and restoring important ecosystem services; and promote co-ordination of biodiversity conservation in the Lake Erie basin.	87 agencies and organizations affecting or affected by Lake Erie, including The Nature Conservancy, Environment Canada, Nature Conservancy of Canada, U.S. Environmental Protection Agency, Michigan Natural Features Inventory, Ohio Lake Erie Commission, Ontario Ministry of Natural Resources, U.S. Fish and Wildlife Service, etc.

plants and animals. But it wasn't always wetlands. Aerial photographs from the 1940s and 1950s show that at least a portion of the property had been farmed. Then in the early-1990s, Waste Management of Michigan, Inc. purchased the property to mitigate for loss of 25.9 ha (64 acres) of wetlands at a landfill that was being expanded in Wayne, Michigan (Hartig, 2008). This wetland mitigation was overseen by the Michigan Department of Natural Resources and a "conservation easement" was placed on the property in 1992 to protect these wetlands in perpetuity. No development can ever occur on this property.

All wetland restoration on site was completed in 1993. In total, 42.9 ha (106 acres) of new wetlands were created on site (ASTI Environmental, 2006). Along with 58.7 ha (145 acres) of pre-existing wetlands, this site now had a total of 101.6 ha (251 acres) of wetlands and 43.3 ha (107 acres) of upland habitats (ASTI Environmental, 2006). The state permit for wetland restoration required a minimum of five years of mitigation monitoring. Based on the five years of monitoring, some additional improvements to the dike and spillway structures had to be completed in 2004 to promote and sustain wetland hydrology.

The state wetland permit also required that monitoring data confirm that these wetlands were functioning as designed. Following scientific confirmation that these wetlands were indeed functioning as designed, Waste Management of Michigan, Inc. transferred this 144.9-ha (358-acre) wetland jewel to the U.S. Fish and Wildlife

Service in 2009 to be incorporated into the DRIWR. Today, these wetlands provide outstanding stopover habitat for many species of birds on their continental migrations and aid in flood control, groundwater recharge, and nonpoint source pollution control. They also provide a "living laboratory" for students from Gibraltar Carlson High School and Shumate Middle School. These students can literally walk out their school door and get an exceptional outdoor environmental educational experience. Further, the site has a local greenway trail that runs along the southern edge of the property that connects to 80 km (50 miles) of continuous greenways trails and the Gibraltar Duck Hunters Association has placed Wood Duck (*Aix sponsa*) nesting boxes in the wetlands. It is now called the Gibraltar Wetlands Unit of the DRIWR and is another excellent example of a public-private partnership to protect natural capital, conserve wildlife, enhance environmental education in local schools, improve quality of life, and strengthen a growing ecotourism economy.

Just north of the mouth of the River Raisin in Monroe, Michigan sits a 97-ha (242-acre) coastal wetland that is today called Ford Marsh (Figure 13). On its northern side is Sterling State Park and on its southern side is the mouth of the River Raisin. The eastern side of the marsh has a beautiful sand beach on western Lake Erie.

Native Americans long knew of the natural resource bounty this region provided. When Jesuit missionaries and Coureur des Bois settled Monroe in 1620, they too marveled at the plethora of natural resources, describing them as a "nature lover's dream" and a sportsman's "perfect paradise" (Bulkley, 1913). What drew Native American and French settlers to this area were the extensive marshes that served as an incredible staging area for waterfowl and as an incredible spawning and nursery grounds for fishes. Can you imagine these early hunters and fishermen looking out over extensive beds of Wild Rice (*Zizania aquatica*) and Wild Celery, and seeing incredible numbers of waterfowl? At the same time they undoubtedly recognized that these remarkable coastal wetlands would support an exceptional fishery. Can you picture the excitement in their eyes?

In its early days, the mouth of the River Raisin was shallow as it meandered its way through the marshland and many sandy shoals. Seeing the effect the Erie Canal had in New York, Monroe city fathers petitioned Congress in 1834 to construct a navigational canal 30.5 m (100 feet) wide, 1,219 m (4,000 feet) long, and 3.7 m (12 feet) deep from the mouth of the River Raisin to western Lake Erie. The canal was dug in various stages and completed in 1843. The new Monroe Harbor had great commercial importance in the State of Michigan because it was the only Michigan port on Lake Erie and it would soon be connected with rail lines.

During this same time period Monroe residents were drawn to the lake for outdoor recreation. Two famous sportsmen's clubs were organized to hunt along the lakefront at what became known as the Monroe marshes: the Golo Club organized in 1854; and the Monroe March Club organized in 1881 (after the closure of the Golo Club). These clubs had a national reputation for world-class hunting and for meeting the culinary desires of wealthy businessmen (Heywood, 1901). Punters, a chef, and servants relieved club members of the drudgery of a hunter's life (Heywood, 1901). To provide a perspective on the quality of the hunting, over 3,000 birds were taken during the 1865 hunting season at the Golo Club, with a daily average of 40 birds per gun (Heywood, 1901). The largest number of birds brought in by a single gun was 157 in the spring of 1883, 145 being Lesser Scaup (*Aythya affinis*) and Ring-necked Duck (*Aythya collaris*)(Wing, 1890).

By the 1890s, the Monroe Harbor waterfront was experiencing large numbers of summer visitors. Large passenger boats brought tourists from far and wide. Several Monroe-owned passenger boats, carrying up to 500 passengers, were bringing tourists to the harbor and its Monroe marshes (Bulkley, 1913). To accommodate these visitors they even built a hotel on the beach in 1895 called "Hotel Lotus" (Higgins, 2000). It was known for its excellent menu and had a bathhouse with 300 changing rooms for those who came to swim at one of Lake Erie's finest beaches. This resort hotel became famous and brought tourists from Detroit, Toledo, Sandusky, Cleveland, and Canada. An electric trolley line was even installed from downtown Monroe to the waterfront and what local residents called the Monroe Piers. This trolley was linked to an extensive electric train network called the "Interurban" that connected Monroe to Detroit, Toledo, Cleveland, Jackson, Lansing, Chicago, and more.

It was easy to see why the Monroe Piers became a huge attraction, with the adjacent Monroe marshes, its famous beaches, Hotel Lotus, the Monroe Yacht Club, a "Casino Building" with a dance pavilion,

Figure 13. Ford Marsh Unit of the Detroit River International Wildlife Refuge (photo credit: ACH).

and a roller coaster with a 22.9-m (75-foot) drop right on the beach. In the summer of 1905 alone, 70,000 passengers were carried on the "Beach Line" trolley to the Monroe Piers (Higgins, 2000).

The 1920s, however, saw the end of the golden era for the Monroe Piers (Higgins, 2000). The growing popularity of the automobile resulted in more people traveling farther afield on the ever-improving roads. Faced with a serious loss of revenue, the Monroe Piers sold its 141 ha (350 acres) to the Newton Steel Company in 1929. With the stock market crash of 1929, the resort never reopened and the buildings were torn down. This began the industrial era.

The land adjacent to the marsh was ideal for industrial manufacturing because of its water access to Lake Erie and the St. Lawrence Seaway, and its proximity to rail lines and a major interstate highway. Newton Steel opened a manufacturing plant on this site in 1929. This plant was purchased by Alcoa in 1942, Kelsey-Hayes in 1947, the Ford Motor Company in 1949, Visteon in 2000, and finally Automotive Components Holdings, Inc. in 2005. Most recently, this manufacturing facility was an automobile parts supplier, including coil springs, stabilizer bars, catalytic converters, and other chassis-related products. Operations ceased in 2008.

Today, Ford Marsh is the largest remaining portion of the Monroe Marshes that was enjoyed and appreciated by so many hunters and tourists during the 1800s. It was given its name in honor of Henry Ford, II who had a hunting cabin on the marsh when it was owned by Ford Motor Company. Despite considerable industrial operations adjacent to the marsh, no industrial activity ever occurred at or in Ford Marsh. It is still well known for significant beds of the American Lotus (*Nelumbo lutea*), a pale yellow flower that is the nation's largest aquatic wildflower. Today, the American Lotus is a state "threatened species" and the official symbol for clean water in Michigan. Interestingly, the Monroe Garden Club hosts an annual American Lotus Tour that makes a stop at Ford Marsh. Ford Marsh remains well known as a stopover site for migratory birds. Local residents continue to recognize Ford Marsh for its diverse wildlife, much like outdoorsmen and tourists did in the 1800s.

In 2005, Automotive Components Holdings entered into a cooperative management agreement with U.S. Fish and Wildlife Service for conservation of Ford Marsh as part of the DRIWR. After five years of cooperative management, ACH and Ford Motor Company donated Ford Marsh to the refuge in 2010. Those early hunters, fishermen, and tourists that so appreciated the Monroe marshes would indeed be so proud of the public-private partnership that has now preserved Ford Marsh in perpetuity as part of the DRIWR, reclaiming its national and international reputation as an incredible "nursery of life," and, once again becoming a major source of community pride.

Land Purchases

One of the early Canadian conservation successes was the purchase of Peche Isle, a 35-hectare (86.5 acre) island located at the head of the Detroit River. In 1883, it was purchased by Hiram Walker, the founder of the famous Hiram Walker Distillery in Windsor, Ontario, for his summer estate that included a large, 54-room home, stable, greenhouse, and icehouse. The property was sold in 1907 to the Detroit, Belle Isle & Windsor Ferry Company. The owner of the Detroit, Belle Isle & Windsor Ferry Company lived in the Walker estate while he unsuccessfully tried to develop the island into an amusement park. Over the years, island ownership changed hands several times with the intent of developing Peche Isle either for exclusive residential development or an amusement park. In the 1960s there was even one developer who completed some work that included a few buildings, with sewage, water, electricity, and telephone connections to the mainland. However, this exclusive resort community never materialized due to mismanagement, lack of funding, and local opposition. In the 1970s, the island was eventually acquired by the Province of Ontario for a nature area, complete with groomed trails, mooring facilities, and shelters. However, this provincial plan never fully materialized due to the lack of funding.

Since the 1970s, local residents have been concerned about the possibility of the island being purchased for private development. These concerns were even raised at the 1998 Canada-U.S. Conference titled "Rehabilitating and Conserving Detroit River Habitats" (Tulen et al., 1998). Based on this citizen concern for possible private development of Peche Isle, this conference concluded that there was an urgent need to protect the few remaining natural areas along the Detroit River, like Peche Isle, and recommended conservation of this unique island. Based on strong local support for conservation and the recommendation of this Canada-U.S. conference, the City of Windsor purchased Peche Isle from the Province of Ontario in 1999 for $1 million. It is now protected in perpetuity for conservation and outdoor recreation, and has been placed on the Canadian registry of lands for the DRIWR. Each year, the Detroit River Canadian Cleanup committee hosts a *Peche Island Day* to explore the tranquility and beauty this island, and to promote its conservation. Organizers of *Peche Island Day* include: Citizens Environment Alliance, BASF, Province of Ontario, Windsor-Essex County Canoe Club, Essex County Field Naturalists' Club, the City of Windsor, and the Detroit River Canadian Cleanup. A unique partnership with Essex County Field Naturalists' Club and the City of Windsor even built a successful nesting platform for Bald Eagles that can be seen from the mainland.

Typically, in large urban areas the price of land is expensive because of the number of people, high demand for land, particularly waterfront lands, and high fair-market property values. Therefore, funding partnerships are often essential to help purchase lands for conservation purposes.

On the U.S. side, as refuge legitimacy was established and trust was built among stakeholder groups, opportunities opened to establish funding partnerships for refuge land purchase. Many industries, foundations, and nonprofit organizations have partnered with the DRIWR to acquire high quality U.S. lands. Table 7 provides three examples of U.S. land purchases using creative public-private partnerships. In each case, industries and/or nonprofit organizations played critical roles in providing funding and facilitating acquisition. U.S. Fish and Wildlife Service could not have acquired these lands without these public-private partnerships.

One of the important lessons learned was not that high quality land was purchased to grow the refuge, but how it was purchased through innovative partnerships. Key success factors included:

- effective working relationships;
- trust among partners;
- timely delivery by a core delivery team;
- well recognized benefits to all stakeholders;
- high profile champion like U.S. Congressman John Dingell; and
- public celebration of successes.

Cooperative Management Agreements

As noted above, the refuge's enabling legislation calls for the use partnerships in undertaking conservation initiatives and states that there will be no forced acquisitions or takings. Further, the refuge's

Table 7. Selected examples of Refuge land acquisitions undertaken using innovative public-private partnerships.

Element	Calf Island Unit	Refuge Gateway	Humbug Marsh Unit
Property Owner	U.S. Fish and Wildlife Service	Wayne County	U.S. Fish and Wildlife Service
Size	4 ha (11.4 acres)	17 ha (44 acres)	165 ha (410 acres)
Location	Grosse Ile Township, Michigan	Trenton, Michigan	Trenton and Gibraltar, Michigan
Significance	Part of the "conservation crescent" that is an archipelago of islands and marshes surrounding the southern end of Grosse Ile at the mouth of the Detroit River	A former industrial brownfield that was cleaned up and restored as an ecological buffer for Humbug Marsh Unit and the future home of the refuge's visitor center	Michigan's only "Wetland of International Importance" designated under the Ramsar Convention
Total Cost	$125,000	$2.5 million	$4.13 million
Funding Partners	Purchased with North American Wetland Conservation Act funding; match funding provided by numerous partners, including Solutia, Ducks Unlimited, and Waterfowl USA	National Oceanic and Atmospheric Administration - $1 million; Chrysler Corporation - $1.5 million (tax write off)	Migratory Bird Conservation Commission; The Charles Stewart Mott Foundation; General Motors Corporation; Ford Motor Company; Trust for Public Land
Acquisition Date	2002	2002	2004
Lead Acquisition Organization	The Nature Conservancy	Wayne County Parks	Trust for Public Land

Comprehensive Conservation Plan calls for cooperative management. In this cooperative conservation model, the refuge grows primarily through cooperative management agreements with industries, government agencies, and other organizations.

Cooperative management agreements are entered into voluntarily between the U.S. Fish and Wildlife Service and an industry or other land owner. This agreement typically remains in force for 50 years, but can be dissolved with a 30-day notice. All management actions are voluntary and at the consent of both parties. The lands (including wetlands and other aquatic resources) become a management unit of the DRIWR and are managed for conservation of wildlife and for wildlife-compatible uses such as environmental education, research, wildlife observation, interpretation, etc., as appropriate. Both parties are indemnified from liability.

Examples of land owner benefits of entering into a cooperative management agreement include:

- receiving positive public relations for becoming a partner in building North America's only international wildlife refuge;
- leveraging human and financial resources for wildlife management actions and/or environmental education;
- receiving a higher priority for monitoring and research; and
- becoming a partner on binational projects that improve outdoor recreation and enhance quality of life.

Table 8 presents five examples of cooperative management agreements that have grown the refuge and that help protect wildlife habitat in this continentally-significant fish and wildlife migration corridor. In each case, these partnerships resulted in protecting wildlife and conserving habitats that otherwise would not have been possible.

Wetland Restoration

Situated at the mouth of Swan Creek in Berlin Township of Monroe County, Michigan is a 61-ha (152-acre) "ecological jewel" in the necklace of refuge lands that span the Detroit River-western Lake Erie corridor. It was the first property purchased by the U.S. Fish and Wildlife Service for the DRIWR in 2003. When purchased, it included over 32 ha (80 acres) of Great Lakes coastal wetlands and 28 ha (70 acres) of agricultural land. The agricultural land had been farmed for over 100 years. To farm this land they had to artificially drain the fields with tiles and ditches, and periodically pump water into western Lake Erie. It is now named the Brancheau Unit after the family that historically farmed the land.

In 2005, U.S. Fish and Wildlife Service and Ducks Unlimited joined forces to develop a plan to restore coastal wetlands in the agricultural fields using dikes, three water control structures (to be able to emulate natural conditions that historically existed on site), and a pump. The project was completed in 2009, restoring wetland functions to 27 ha (67 acres) of farmland. Of remarkable interest was the immediate wetland response. No seeding occurred with any wetland species, yet cattails, arrowhead, sedges, and other aquatic plants immediately sprouted. This seed base had remained viable in the agricultural land for over 100 years (Figure 14).

This wetland restoration was particularly significant because over 90% of the historical wetlands in the western Lake Erie watershed had been lost to agricultural, municipal, and commercial development. The project redirected most of the runoff that came from the local watershed into the restored wetlands, allowing natural wetland processes to clean the water before it is discharged to western Lake

Figure 14. Restoration of 27 ha of wetlands on former agricultural land at the Brancheau Unit of the Detroit River International Wildlife Refuge (photo credit: U.S. Fish and Wildlife Service).

Table 8. Selected examples of cooperative management agreements (with industries and nongovernmental organizations) that have been used by the U.S. Fish and Wildlife Service to add lands to the Detroit River International Wildlife Refuge.

Characteristic	Lake Erie Metropark Unit	Lagoona Beach Unit	Lady of the Lake Unit	Erie Marsh Unit	Gard Island Unit
Location	Brownstown Township, Wayne County, Michigan	Frenchtown Township, Monroe County, Michigan	Erie Township, Monroe County, Michigan	Erie Township, Monroe County, Michigan	Erie Township, Monroe County, Michigan
Size	315 ha (780 acres)	265 ha (656 acres)	19 ha (49 acres)	897 ha (2,217 aces)	7 ha (18 acres)
Property Owner	Huron Clinton Metropolitan Authority	DTE Energy	Consumers Energy	The Nature Conservancy	The University of Toledo
Geography and Ecology	A unique complex of coastal marshes, meadows, lagoons, and other wetland habitats located at the mouth of the Detroit River as it enters western Lake Erie	A unique complex of upland habitats, coastal wetlands, and prairie habitats located south of the mouth of Swan Creek on western Lake Erie	731.5 m (2,400 feet) of sand beach on western Lake Erie; two natural, land-locked coastal wetlands separated from Lake Erie by a sand beach ridge	Represents 11% of the remaining marshland in Michigan waters of western Lake Erie	Part of Erie Marsh; a unique coastal island that helps support ecosystem integrity of Erie Marsh
Natural Resources	Unique biodiversity; provides critical stopover habitat for migratory birds, including waterfowl, shorebirds, landbirds, etc.	Unique biodiversity; provides critical stopover habitat for migratory birds, including waterfowl, shorebirds, landbirds, etc.	Unique biodiversity; provides critical stopover habitat for migratory birds, including waterfowl, shorebirds, landbirds, etc.	Exceptional biodiversity; provides critical stopover habitat for migratory birds, including waterfowl, shorebirds, landbirds, etc.	Provides critical stopover habitat for migratory birds, including waterfowl, shorebirds, landbirds, etc.

Table 8. Cont'd

Public Uses	Part of Lake Erie Metropark; a 80-ha (200-acre) Nature Study Area located on site; considered one of the three best places in the U.S. to watch Hawk migrations; hosts Detroit River Hawk Watch program that tracks raptor migrations using citizen science; sponsors Hawkfest each September that attracts over 7,000 people	No public access or public use because of restrictions pertaining to the nuclear power plant	Public access is provided to the beach area; unit well known for fishing and birding in the area	Well known by birders for exceptional birding; Erie Shores Birding Association hosts birding events on site	University of Toledo uses the island for environmental education and research; community groups perform annual cleanups
Other partners	Numerous partners support Hawkfest and Detroit River Hawk Watch, including International Wildlife Refuge Alliance, Osprey Watch of Southeast Michigan, Detroit Audubon Society, etc.	Habitat restoration partners have included Ducks Unlimited, Waterfowl USA-Southwestern Lake Erie Chapter, Pheasants Forever, and Metropolitan Affairs Coalition	Local school groups do annual cleanups of the beach	Erie Fishing & Shooting Club was the original owner of the property dating back to the 1800s and still retains a lease for hunting privileges	Local community groups do annual cleanups of the island

Erie. This resulted in reducing nonpoint source loadings of herbicides, pesticides, and nutrients to western Lake Erie. This project also greatly reduced the risk of flooding to local residences, businesses, and agricultural parcels.

Today, U.S. Fish and Wildlife Service manages water levels in these wetlands using the pump, the three water control structures, and a flap gate. These water control structures provide flexibility in the management of the wetlands, allowing the U.S. Fish and Wildlife Service to manage for a variety of wildlife species, as well as provide management capabilities to combat invasive species such as *Phragmites*. The structures also prevent unwanted water from backing up into the unit and adjacent properties during a wind event. Once wetland functions returned, water levels were manipulated to achieve a desired plant response according to a water level management plan.

Finally, the project resulted in increasing outdoor recreational opportunities for local residents. In 2012, it was opened to waterfowl hunting as part of the Refuge's hunt program. Hunter response and hunting success have been very favorable. This innovative project could not have happened without a number of public and private partners, including U.S. Fish and Wildlife Service, Ducks Unlimited, Waterfowl USA, the International Wildlife Refuge Alliance, Michigan Duck Hunters Association, Metropolitan Affairs Coalition, DTE Energy, and the North American Wetlands Conservation Council.

Invasive Weed Management

Invasive species are a major threat to the ecological integrity of ecosystems worldwide and billions of dollars are spent on their control. The Common Reed (*Phragmites australis*) is an invasive plant that has become dominant in many southeast Michigan and southwest Ontario wetlands, including the Refuge's coastal wetlands. One form of common reed is native to North America, but another form was introduced from Europe and is much more aggressive than the native variety. *Phragmites* grows very fast and can reach 4.6 m (15 feet) in height, forming a dense canopy. It literally outcompetes other native wetland plants and alters wetland hydrology. The dense canopy and duff layer reduces light availability for more beneficial native plants.

Because *Phragmites* is ubiquitous throughout the region and Refuge, and there are many landowners impacted, an innovative partnership was formed to control *Phragmites* and other invasive plants on a broad geographic scale. The result was a partnership among governments, industries, and nongovernmental organizations to accomplish conservation results collectively that could not be accomplished individually. BASF Corporation, DTE Energy, Ducks Unlimited, Eastern Michigan University, Huron-Clinton Metropolitan Authority, International Wildlife Refuge Alliance, Michigan Department of Natural Resources, Monroe Conservation District, Southeast Michigan Council of Governments, The Stewardship Network, The Nature Conservancy, Wildlife Habitat Council, and U.S. Fish and Wildlife Service partnered to create the Detroit River-Western Lake Erie Cooperative Weed Management Area. All parties signed a Memorandum of Understanding and pledged to partner on:

- inventory and monitoring;
- prevention of establishment of new invasive plant species;
- invasive species management (e.g. herbicide treatment, cutting, prescribed fires);
- cooperation on funding, staff training, equipment sharing, etc.;
- information exchange; and
- broad-based educational efforts.

A formal steering committee oversees all work and a *Phragmites* Strike Team was established to achieve on-the-ground conservation results. These partners have been successful in: obtaining grant funding to control *Phragmites* on a corridor basis; obtaining a Marsh Master (i.e. amphibious vehicle) to cut and treat *Phragmites*; herbicide-treating over 485 ha (1,200 acres) of wetlands owned by the partners and adjacent property owners in 2012 alone; and conducting prescribed fires. For example, 89 ha (220 acres) were burned at the Strong Unit and adjacent Great Lakes Aggregates property in 2012 and 22 ha (55 acres) were burned at the Fix Unit in 2013. These prescribed fires were undertaken to help revitalize native plants within wet meadows, wet prairies, and forested wetland areas. Clearly, this is an outstanding example of many public and private partners working through a local organization (i.e. Cooperative Weed Management Area) to integrate all invasive plant management resources across jurisdictional boundaries to benefit a broader geographic region.

Public Use Improvements

Public-private partnerships have also been instrumental in furthering public use in the refuge. The Naval Construction Battalion 26 of the Navy Seabees needed a training exercise and approached the U.S. Fish and Wildlife Service about potential construction projects that could be used as a training exercise. In 2008, these Navy Seabees removed over 1,125 m (0.7 mile) of old concrete road in Humbug Marsh, crushed the concrete to the desired engineering size (i.e. 21 aa), and used it build the base of approximately 800 m (0.5 mile) of universally-accessible trails. The Navy Seabees and numerous partners also constructed approximately 1,600 m (one mile) of rustic trails, a wetland boardwalk, a pedestrian stream crossing connecting Humbug Marsh Unit with the Refuge Gateway, and an environmental education shelter in Humbug Marsh. Other key partners in the project included NTH Consultants, DTE Energy, Mid-American Group, Lowe's Home Improvement, and many more. In total, over 40 businesses and organizations contributed equipment, money, or in-kind services to complete this project. The total value of the project was over $500,000, with no financial resources provided by U.S. Fish and Wildlife Service. The end result for the refuge was that it could now provide an outstanding environmental education and interpretation experience at the Refuge Gateway and Humbug Marsh Unit – Michigan's only "Wetland of International Importance" designated under the international Ramsar Convention (Figure 15).

Concluding Thoughts

Urban areas are characterized by high human population densities, institutional complexity, and often limited open and green spaces. This makes bringing conservation to such places particularly challenging. Therefore, partnerships are essential. One good way of thinking about conservation partnerships in urban areas is to visualize it through the analogy of farming that includes "preparing the ground," "sowing the seed," "tending the fields," and "reaping the harvest" (World Health Organization, 2003). There is no standard blueprint for building conservation partnerships. The process of building and sustaining partnerships must be tailored to the particular urban area and be flexible, not rigid.

Figure 15. Celebration of the completion of construction of an environmental education shelter, 2,400 m of trails, a wetland boardwalk, and a pedestrian stream crossing in Humbug Marsh Unit in 2008 (photo credit: D. Mitchell).

In the "preparing the ground" phase of a partnership, it is important to bring potential partners together and reach agreement on mutual needs, a common vision, and shared values. Once the partnership comes together around shared needs and a common vision, it can move onto the next phase of "sowing the seeds." This involves reviewing project options and reaching agreement on a preferred project option or options to be implemented. As part of this phase it is also important to reach agreement on priorities and partner roles and responsibilities for implementation. In the next phase of "tending the fields" partners coordinate implementation of the project or projects within a consensus timeframe and undertake mid-course corrections, as necessary, to achieve desired outcomes. In the final phase of "reaping the harvest" partners perform assessments to confirm that desired outcomes have been achieved and celebrate this achievement in a very public fashion to ensure partner satisfaction.

Although there has been some success in transboundary conservation since the 2001 Conservation Vision, much remains to be done to fully incorporate an ecosystem approach in natural resource and environmental decision-making on a binational watershed scale. Clearly, the public-private partnership examples presented above show the potential for achieving significant conservation results in a major urban area. Further, the public-private partnership experiences of the DRIWR have shown that urban conservation initiatives can help enhance urban identity, promote sustainability, improve community pride, and help achieve urban competitive advantage. This can also help establish conservation relevance in a growing urban population. Critical to this is fostering a more informed citizenry that actively supports and understands the value of conservation.

Literature Cited

ASTI Environmental, 2006. Gibraltar Wetland Creation /Restoration, Mitigation Wetland Monitoring Report No. 5, City of Gibraltar, Wayne County, Michigan. Waste Management of Michigan, Inc. Wayne, Michigan, USA.

Bulkley, J.M., 1913. *History of Monroe County, Michigan*. The Lewis Publishing Company, Chicago, Illinois, USA.

Essex Region Conservation Authority, 2002. Essex Region Biodiversity Conservation Strategy - Habitat Restoration and Enhancement Guidelines (Comprehensive Version). Essex, Ontario, Canada.

Hartig, J., 2008. Environmental Site Assessment of the Waste Management Tract, Gibraltar, Michigan. U.S.Fish and Wildlife Service, Grosse Ile, Michigan, USA.

Heywood, F., 1901. Two famous sportmens' clubs. Field and Stream. May Issue (No. 3), 131-137.

Higgins, M., 2000. Monroe's Lighthouses: Lost in the pages of time. In: J. Harrison (Ed.), *Lighthouse Digest*. Foghorn Publishing, East Machias, Maine, USA. (April 2013).

MacKenzie, S.H., 1996. *Integrated Resource Planning and Management: An Ecosystem Approach in the Great Lakes Basin*. Island Press Publishing, Washington, D.C., USA.

Nolan, J., 1999. How Prohibition made Detroit a bootleggers dream town. The Detroit News, June 15th Edition, Detroit, Michigan, USA.

Pearsall, D., Cavalieri, C., Chu, C., Doran, P., Elbing, L., Ewert, D., Hall, K., Herbert, M., Khoury, M., Kraus, D., Mysorekar, S., Paskus, J., Sasson, A., 2012. Returning to a Healthy Lake: Lake Erie Biodiversity Conservation Strategy. Technical Report. A joint publication of The Nature Conservancy, Nature

Conservancy of Canada, and Michigan Natural Features Inventory. Lansing, Michigan, USA.

Tulen, L.A., Hartig, J.H., Dolan, D.M., Ciborowski, J.J.H., (Eds.), 1998. Rehabilitating and conserving Detroit River habitats. Great Lakes Institute for Environmental Research Occasional Publication No. 1. Windsor, Ontario, Canada.

U.S. Fish and Wildlife Service, 2005. *Comprehensive Conservation Plan and Environmental Assessment for the Detroit River International Wildlife Refuge.* Grosse Ile, Michigan, USA. http://www.fws.gov/midwest/planning/detroitriver/ (April 2013).

Wing, T. E., 1890. *History of Monroe County, Michigan.* Munsell & Company, Publishers, New York, New York, USA.

World Health Organization, 2003. *A Pocket Guide to Building Partnerships.* Geneva, Switzerland.

Chapter 5

Creating a New Waterfront Porch for People and Wildlife

It was early morning when our boat left the Bay View Yacht Club on Conner Creek in Detroit. As we headed downstream to the confluence with the Detroit River, we were greeted with a spectacular view of Lake St. Clair, Peche Isle at the head of the Detroit River on the Canadian side, and Belle Isle – the 396.6-hectare (980-acre) crown jewel of Detroit's park system initially designed by world-renowned landscape architect Frederick Law Olmstead. We were truly awestruck at how beautiful the river was. The weather conditions were fair and calm, and the surface of the water was like a mirror reflecting distant images of the islands and shorelines. To me it was a nautical canvas, even more beautiful than the water and wildlife paintings of John James Audubon.

Upon entering the Detroit River we headed west and downstream toward downtown Detroit and Windsor, Ontario, Canada, the sister city of Detroit. Because it was early morning it was relatively quiet, with only the sounds of a few Ringed-billed Gulls (*Sterna hirundo*) and our engine. It was so incredibly peaceful and truly captured the beauty of the Detroit River. But as we headed downstream along the U.S. shoreline it was obvious that we were in a major urban area with a long history of waterfront development for commerce and maritime industry. Kilometer after kilometer we passed hardened shorelines made up of concrete breakwaters, steel sheet piling, or broken concrete with rebar protruding in many directions.

I couldn't help but think about the shoreline from a fish and wildlife habitat perspective. Where do the fish and wildlife rest and feed? What happened to the spawning and nursery grounds? Where along this hardened shoreline can the invertebrates live? What have we done to the place they call home?

Hard Shoreline Engineering

The Detroit River has a long history of industrial development dating back to ship building and maritime operations. Indeed, it was common practice in the 20th Century to utilize hard shoreline engineering as part of commercial shoreline development. Hard shoreline engineering is generally defined as the use of concrete breakwaters or steel sheet piling to:

- stabilize shorelines for protection from flooding and erosion;
- achieve greater human safety; and
- accommodate commercial navigation or industry (Caulk et al., 2000).

Although hard shoreline engineering can achieve commercial, navigational, and industrial benefits, it typically results in negative ecological impacts because it provides no habitat and often restricts access to adjacent habitats. Such anthropogenic hardening of shorelines not only destroys natural features and biological communities, but it also alters the transport of sediment, disrupting the balance of accretion and erosion of materials carried along the shoreline by wave action and long-shore currents. This disruption of sediment transport processes can intensify the effects of erosion, causing ecological and economic impacts (Schneider et al., 2009).

Along the Detroit River, 49.9 of the 51.5 km of U.S. mainland shoreline have been hardened with concrete, steel, or both, leaving only 1.6 km of natural shoreline (Manny, 2003). This shoreline hardening contributed to a 97% loss of coastal wetland habitats along the Detroit River (Manny, 2007). Schneider et al., (2009) reported that 47.2% and 20.4% of the entire U.S. and Canadian Detroit River and Lake Erie shorelines, respectively, are "highly protected" using hardening techniques. Changes in the extent of hardening or armoring along western Lake Erie's shoreline have been documented by Ohio Department of Natural Resources since the 1870s (Fuller and Gerke, 2005; Livchak and Mackey, 2007). In the 1930s, less than 5% of shoreline in Ottawa and Lucas counties was hardened or armored. As of the 1990s, 78% of the Ottawa County lakeshore and 98% of the Lucas County lakeshore were hardened or armored, representing substantial loss of natural habitat for biota (Fuller and Gerke, 2005; Livchak and Mackey, 2007).

Soft Shoreline Engineering

With the decrease in the Detroit Metropolitan Area's industrial activity (particularly on the waterfront), the expansion of other modes of transportation (i.e. more goods and people being transported by rail, highway, and air), a growing environmental consciousness (i.e. more people being aware of how they are part of an ecosystem and what they do to their ecosystem they do to themselves), and the growing public interest in close-to-home, outdoor recreation, there has been a unique opportunity and growing interest in reclaiming and redeveloping portions of the waterfront for multiple purposes so that additional benefits can be accrued. Indeed, the interest in redeveloping the Detroit River waterfront for additional benefits beyond commerce has been generated by the Southeast Michigan Greenways Initiative that is catalyzing construction of non-motorized trails that link parks, nature areas, cultural features, and historic sites for recreational and conservation purposes (Hartig et al., 2007), and the Detroit River Remedial Action Plan, the blueprint for cleaning up the river and restoring impaired beneficial uses, that called for restoring degraded fish and wildlife habitats in the Detroit River Area of Concern (Canada Ontario Agreement Respecting Great Lakes Water Quality and Michigan Department of Natural Resources, 1991). Further, the Comprehensive Conservation Plan for the Detroit River International Wildlife Refuge calls for the restoration of coastal wetlands and other riparian habitats along the shoreline of the Detroit River and western Lake Erie in an effort to restore and sustain fish and wildlife communities (U.S. Fish and Wildlife Service, 2005).

Out of this convergence of waterfront redevelopment opportunity and societal-environmental interest, came the concept of soft shoreline engineering. Soft shoreline engineering is the use of ecological principles and practices to reduce erosion and achieve stabilization and safety of shorelines, while enhancing riparian habitat, improving aesthetics, and even saving money (Caulk et al., 2000; Hartig et al., 2001). Soft shoreline engineering is achieved by using vegetation and other materials to improve the land-water interface by enhancing ecological features without compromising the engineered integrity of the shoreline.

It should be noted that riprap alone is considered a form of hard shoreline engineering. Riprap is generally understood to be a permanent cover of rock and stone used to stabilize stream banks and lakeshores, and to provide in-stream channel stability. The Reston Association (1996) notes that riprap, under most circumstances, adequately stabilizes shorelines and does not share all the negative characteristics of bulk heading, however, it does not provide optimal wildlife and water quality benefits as vegetated shorelines. Therefore, the Reston Association (1996) recommends that if riprap is employed as a shoreline treatment, that it includes vegetative plantings to improve habitat and provide some ecological and water quality benefits. The use of riprap in conjunction with vegetative plantings is sometimes referred to as the "greening of riprap" or "vegetated riprap" and is considered one of the accepted soft shoreline engineering techniques.

In 1999, a group of U.S. and Canadian researchers and natural resource managers convened a conference on soft shoreline engineering and developed a best management practices manual (Caulk et al., 2000) to encourage and catalyze use of soft shoreline engineering techniques. That best management practices manual was then used to educate people on the value and benefits of soft shoreline engineering, and to recruit communities, developers, industries, and others to incorporate it in redevelopment and rehabilitation efforts.

Think of soft shoreline engineering along the Detroit River as re-creating a new waterfront porch for both people and wildlife (Figure 16A and B). Clearly, there is growing interest among urban planners and developers to reclaim waterfronts with new front porches. The challenge is to work with and through these urban planners and developers to incorporate soft shoreline engineering, where feasible. Clearly, it would not be feasible at a port facility that wanted to off load freighters or bring in cruise ships. But where it is feasible, soft shoreline engineering can help achieve many benefits, including reducing erosion and achieving stability and safety of shorelines, enhancing habitat, improving aesthetics, enhancing urban quality of life, increasing waterfront property values, and even saving money when compared to installing concrete breakwaters or steel sheet piling.

Implementing Soft Shoreline Engineering and Lessons Learned

Since 1998, a total of 53 soft shoreline engineering projects have been implemented in the Detroit River and western Lake Erie watershed (Table 9). Of these 53 projects, 49 were undertaken with multiple

A

B

Figure 16. DTE Energy's Rouge Power Plant shoreline before (A) and after restoration (B) using soft shoreline engineering techniques, River Rouge, Michigan, USA (Photo Credit: Nativescape, LLC).

Table 9. A survey of soft shoreline engineering projects implemented in the Detroit River-western Lake Erie watershed, 1996–2013 (all Michigan project expenditures are reported in U.S. dollars; all Ontario project expenditures are reported in Canadian dollars). More detailed information on these projects is available at: www.stateofthestrait.org.

Location	Project Goals	Project Description and Cost	Time frame	Partners	Monitor-ing
BASF Park, Wyandotte, Michigan	Demonstrate use of Elastocoast (Elastomeric revetment that stabilizes shorelines and enhances habitat by increasing interstitial spaces) along the Detroit River shoreline of BASF Park	Stabilized shoreline to a depth of 37 centimeters with five-centimeter crushed limestone bound together with the Elastocoast product; $6,000	2008	BASF Corporation, City of Wyandotte	Qualitative
BASF Riverview, Trenton Channel, Riverview, Michigan	Remediate a contaminated site, add incidental habitat to steel sheet piling walls, and create one acre of fish spawning habitat	Following remediation of a contaminated site, incidental habitat was added to 366 meters of steel sheet piling, and one acre of Walleye, Smallmouth and Largemouth Bass, and Sturgeon spawning habitat was created; $100,000	2007–2008	BASF Corporation	None
Blue Heron Lagoon (west shoreline) on Belle Isle, Detroit, Michigan	Restore emergent wetland shoreline and enhance wildlife habitat	Controlled invasive species, planted native species in the upland buffer area, and placed logs along shoreline to provide habitat to native turtles; $34,000	2000	Detroit Recreation Department, U.S. Fish and Wildlife Service, Michigan Sea Grant, and seven other partners	Qualitative

Table 9. Cont'd

Blue Heron Lagoon and canals on Belle Isle, Detroit, Michigan	Restore fish access and habitat	Restore access to 38 ha of wetlands and shallow and deep water habitat, and 3.5 km of canal habitat; restore two ha of fish rearing and nursery habitat; $1.5 million	2012-2013	Detroit River Remedial Action Plan, Friends of the Detroit River, JJR, and others	None
Central Waterfront, Windsor, Ontario	Restore 550 m of shoreline using soft shoreline engineering techniques	550 m of unstable and eroding shoreline was stabilized using soft shoreline engineering techniques, including enhancement of shoreline crenulation using a diversity of substrate types and sizes; a 125-m cobble beach was anchored by two large rock promontories provides one of the only river access locations on the Canadian side of the upper Detroit River; $1.3 million	2011	Essex Region Conservation Authority, City of Windsor, Environment Canada, Ontario Ministry of the Environment, and Ontario Ministry of Natural Resources	None to date; fishery monitoring planned in future
Danvers Pond, Farmington Hills, Michigan	Remove existing dam, stabilize shoreline using soft shoreline engineering techniques, and restore natural flows	Demolished existing dam, restore natural fish passage, stabilized 91 m of shoreline using soft shoreline engineering techniques, and restored 0.8 ha of fish and wildlife habitat; $500,000	2010-2012	Alliance of Rouge Communities, City of Farmington Hills, Wayne County, and U.S. Environmental Protection Agency	None to date
Dean Construction Site, LaSalle, Ontario	Naturalize 550 meters of shoreline and create a 0.45-hectare storm water management system to treat runoff	Restored 550 meters of natural shoreline using soft engineering techniques, reestablished 0.55 kilometers of riparian vegetation along the natural shoreline and created a storm water pond to improve the quality of the storm water before it enters the Detroit River; $62,000	1997–1998	Dean Construction, Environment Canada, and Ontario Ministry of Natural Resources	Qualitative

Table 9. Cont'd

Detroit RiverWalk - Stroh River Place, Detroit, Michigan	Build a section of the Detroit RiverWalk in front of Stroh River Place and enhance riparian habitat	Built a 305-meter section of the Detroit RiverWalk using a cantilever design with habitat features beneath the cantilevered RiverWalk; $1 million	2006–2007	Detroit Riverfront Conservancy, Stroh Companies, Inc., Omni Hotel, and Tallon Industries	None
Detroit RiverWalk - West of Milliken State Park, Detroit, Michigan	Stabilize the shoreline along the Detroit RiverWalk and enhance aquatic habitat	Stabilized 152 meters of shoreline with varying sizes of rock armor stone and enhanced aquatic habitat; $100,000	2003–2004	Detroit Riverfront Conservancy and General Motors Corporation	None
Detroit River waterfront (between Lincoln and Langlois Ave.), Windsor, Ontario	Restore 500 meters of shoreline using soft engineering techniques and enhance fish habitat	Converted old vertical seawalls into gently sloping irregular rock shoreline configurations and enhanced fish habitat by planting native species; $70,000	1998	City of Windsor, University of Windsor, Dean Construction, and Ontario Ministry of Natural Resources	Qualitative
DTE's Rouge Power Plant, River Rouge, Michigan	Remove broken concrete and asphalt, stabilize shoreline, and enhance habitat	Reconstructed 61 meters of natural shoreline using soft engineering techniques and reestablished a natural riparian buffer made up of four Michigan native plant communities; $30,000	2005	DTE Energy, Nativescape, U.S. Fish and Wildlife Service, Department of Environmental Quality, Michigan Sea Grant, and five other partners	Qualitative

Table 9. Cont'd

DTE's Rouge Power Plant, River Rouge, Michigan	Remove broken concrete and asphalt, stabilize shoreline, and enhance habitat	Reconstructed 61 meters of natural shoreline using soft engineering techniques and reestablished a natural riparian buffer made up of four Michigan native plant communities; $30,000	2005	DTE Energy, Nativescape, U.S. Fish and Wildlife Service, Department of Environmental Quality, Michigan Sea Grant, and five other partners	Qualitative
DTE's Monroe Power Plant, Monroe, Michigan	Restore 152 meters of natural shoreline and enhance fish and migratory bird habitat	Restored 152 linear meters of the River Raisin shoreline, created a wetland edge and a five-meter-wide upland buffer area where native species were planted; $68,000	2007–2008	Metropolitan Affairs Coalition, City of Monroe, U.S. Fish and Wildlife Service, Michigan Department of Environmental Quality, International Wildlife Refuge Alliance, and eight other partners	Qualitative
Ellias Cove, Trenton, Michigan	Remediate mercury, lead, zinc and PCB contaminated sediment from Ellias Cove and restore the shoreline using soft engineering techniques	Removed 88,000 cubic meters of sediment and disposed contaminated sediment in special contaminant cell at Pointe Mouillee Confined Disposal Facility in western Lake Erie and restored shoreline habitat, including nursery habitat for fish; $150,000 for habitat portion	2006	U.S. Environmental Protection Agency, Michigan Department of Environmental Quality, Great Lakes Basin Program for Soil Erosion and Sediment Control, and seven other partners	Qualitative
Elizabeth Park Canal Shoreline, Trenton, Michigan	Restore natural shoreline using soft engineering techniques, rehabilitate wildlife habitat and improve water quality in canal	Restored a natural shoreline using soft engineering techniques; reduced erosion and runoff with creation of a buffer zone of native trees, shrubs, wildflowers, and grasses; and enhanced fish and wildlife habitat; $40,000	2007–2008	Wayne County Parks, Nativescape, Michigan Sea Grant, U.S. Fish and Wildlife Service, and International Wildlife Refuge Alliance	Qualitative

Table 9. Cont'd

Elizabeth Park –Phase 1, North River Walk, Trenton, Michigan	Stabilize and enhance 183 meters of shoreline and enhance underwater fish habitat	Removed a 1910 concrete breakwall from the north end of Elizabeth Park, stabilized the shoreline using soft engineering techniques, and created two oxbow islands for nursery habitat for fish; $1 million	2001	Clean Michigan Initiative and Wayne County Parks	Quantitative
Elizabeth Park – Phase 2, North River Walk, Trenton, Michigan	Stabilize the Detroit River shoreline, complete the River Walk, and enhance habitat	Removed the deteriorated 1910 breakwall remaining along the shoreline, graded back the shoreline and stabilized it with Armor stone and landscape plantings, and completed River Walk; $1 million	2012	Michigan Natural Resources Trust Fund and Wayne County Parks	None to date
Fort Malden Shoreline, Amherstburg, Ontario	Stabilize shoreline and enhance fish habitat by constructing offshore Lake Sturgeon spawning habitats	Stabilized 300 meters of shoreline, constructed an armor rock revetment and offshore deepwater rock/cobble shoals to enhance fish habitat and create Lake Sturgeon spawning habitats; $290,000	2004	Essex Region Conservation Authority and Parks Canada	Quantitative
Frank and Poet Drain, Trenton, Michigan	Streambed, bank, and upland habitat restoration	Excavated and stabilized shoreline, planted emergent wetland plants and created an upland buffer area with wildflowers and prairie grasses; $80,000	2007–2009	Friends of the Detroit River, National Fish and Wildlife Foundation, and seven other partners	Qualitative
Gabriel Richard Park, Detroit, Michigan	Stabilize river shoreline, restore fish habitat, and provide aesthetically-pleasing environment	Stabilized and restores 300 m of shoreline with fish habitat components, including the addition of two fishing overlooks; $300,000	2006-2007	Detroit Riverfront Conservancy, Detroit Parks and Recreation, and JJR	None to date

Table 9. Cont'd

Gibraltar Bay, Detroit River, Grosse Ile, Michigan	Restore native plant community and promote education and stewardship	Restored 357 meters of shoreline using biodegradable "soil sock" and clean-composted recycled yard waste to create a new aquatic shelf and planted 1,400 emergent plants; $80,000	Phase 1: 2003; Phase 2: 2004–2005	Grosse Ile Nature and Land Conservancy, Nativescape, and eight other partners	Qualitative
Goose Bay in Windsor, Ontario	Stabilize shoreline and enhance fish habitat	Protected shoreline with riprap and native plantings, and enhanced fish habitat; $205,000	1999–2000	Essex Region Conservation Authority, City of Windsor and Environment Canada's Great Lakes Cleanup Fund	Quantitative
Intrepid Pond at intersection of Intrepid and Meridian at the Commerce Park, Grosse Ile, Michigan	Restore storm water retention basin, create native plant community shoreline, and promote education and stewardship	Removed invasive plant species such as *Phragmites australis* and Eurasian milfoil (*Myriophyllum spicatum*), planted native wetland plants along the shoreline, and created an upland buffer with native bushes and trees; $7,000	2008 - 2010	Grosse Ile Nature & Land Conservancy, Alliance for the Great Lakes Freshwater Future, and Ford Motor Company	Qualitative
Lake Muskoday on Belle Isle, Detroit, Michigan – Phase 1	Control erosion and enhance shoreline habitat	Stabilized shoreline using soft engineering techniques, removed invasive plant species such as *Phragmites australis*, and planted native wetland plants, shoreline plants and seeds; $30,000	2000–2001	Detroit Recreation Department, Greater Detroit American Heritage River Initiative, and five other partners	Qualitative

Table 9. Cont'd

Lake Muskoday on Belle Isle, Detroit, Michigan – Phase 2	Restore shoreline using bio-stabilization techniques, including live brushmattress, live siltation, live fascines, vegetated riprap, live brushlayering, and vegetated mechanically stabilized earth	Stabilized shoreline and enhanced riparian habitat using a variety of bio-stabilization techniques as a demonstration project; $10,000	2002	University of Michigan, Detroit Recreation Department, and volunteers	No post-project monitoring, except for photographs
Little River at Twin Oaks, Windsor, Ontario	Stabilize 1,150 meters of shoreline, reestablish the natural floodplain, and reestablish the riparian vegetation to improve fish and wildlife habitat	Created a "Natural Channel Design" which stabilized the natural floodplain, planted riparian native species and placed granular stone at bottom of the meandering stream to improve habitat for fish; $1 million	1997–1998	City of Windsor, Essex Region Conservation Authority, Environment Canada's Great Lakes Cleanup Fund, University of Windsor, and five other partners	Qualitative
Maheras Gentry Park, Detroit, Michigan	Create an oxbow and restore fish and wetland habitat as mitigation for the construction of Conner Creek Combined Sewer Overflow control facility	Removed 38,300 cubic meters of soil for an oxbow, planted native vegetation to improve fish habitat, and created fish spawning and nursery areas; $2.3 million	2000–2004	Detroit Water and Sewerage Department and Detroit Parks and Recreation	Qualitative
McKee Park, Windsor, Ontario	Enhance shoreline habitat and submerged fish habitat for Lake Sturgeon and other species	Protected 182 meters of natural shoreline by constructing offshore barriers using large and small quarry rock to reduce high energy currents and to improve spawning and nursery habitat for fish; $182,000	2003	Essex Region Conservation Authority, City of Windsor, University of Windsor, and eight other partners	Quantitative

Table 9. Cont'd

Mill Race Village, Northville, Michigan	Restore streambank buffer habitat to reduce erosion and improve water quality	Installed 11 coir logs at the base of the slope to stabilize 33.5 m of shoreline, removed invasive plants from a 91.4 m reach of shoreline, and installed an erosion control blanket and native plants to cover approximately 46.5 m^2 of bank; $1,600	2012	Friends of the Rouge and City of Northville	None to date
Mt. Elliott Park, Detroit, Michigan	Restore shoreline using soft shoreline engineering techniques, including stabilizing the land-water interface using limestone of varying sizes and native plant materials in an effort to help restore fish and wildlife habitat	Restored 200 m of shoreline using soft shoreline engineering techniques, including providing an interactive water feature and playscape for children that will creatively teach the importance of water and a Leadership in Energy and Environmental Design (LEED)-certified pavilion; $200,000 for shoreline habitat work	2012-2013	Detroit Riverfront Conservancy and Detroit Recreation Department	None to date
Northeast Shore of Fighting Island, LaSalle, Ontario	Stabilize shoreline and enhance aquatic habitat	Shoreline sinuosity was increased by constructing limestone groynes along the shoreline that increased stability and enhanced habitat; $60,000 (U.S.)	1996	BASF Corporation and Essex Region Conservation Authority	None

Table 9. Cont'd

North Pier Shoreline of Belle Isle, Detroit, Michigan	Remove debris and naturalize the Belle Isle shoreline, including enhancement of coastal wetland habitat	Remove 340 tonnes of non-accumulating debris and naturalize 23 m of shoreline and coastal wetland with native plants and natural rock on Belle Isle to enhance fish habitat; $50,000 (U.S.)	2012-2012	Alliance for the Great Lakes, Great Lakes Restoration Initiative, National Ocean Service's Marine Debris Program, City of Detroit, and Sturgeon for Tomorrow	None to date
Northwest Shore of Fighting Island, LaSalle, Ontario	Demonstrate use of Elastocoast (Elastomeric revetment that stabilizes shorelines and enhances habitat by increasing interstitial spaces) along the Detroit River shoreline of Fighting Island	Stabilized shoreline to a depth of 37 centimeters with five-centimeter crushed limestone bound together with the Elastocoast product; $6,000 (U.S.)	2007	BASF Corporation	Qualitative
Refuge Gateway Shoreline along the Trenton Channel of the Detroit River, Trenton, Michigan	Stabilize shoreline using soft engineering techniques and restore coastal wetland and upland buffer habitats	Stabilized the shoreline using soft shoreline engineering techniques and restored 4.2 ha of emergent marsh, 1.7 ha of submergent marsh, and 4.8 ha of upland buffer habitats; $746,000	2010	Wayne County, Michigan Department of Natural Resources and Environment, U.S. Fish and Wildlife Service, Metropolitan Affairs Coalition and five other partners	Qualitative

Table 9. Cont'd

Refuge Gateway, Monguagon Wetland System Outflow, Trenton, Michigan	Restore 60 m of outflow shoreline using soft shoreline engineering techniques	Used fachines and live stakes to stabilize the shoreline and enhance habitat along the outflow of the Monguagon wetland system that discharges to the Detroit River; $2,500	2011	U.S. Fish and Wildlife Service, International Wildlife Refuge Alliance, Boy Scout Troop # 802, and Downriver Walleye Federation	Qualitative
River Canard Park, Amherstburg, Ontario	Restore 200 m of shoreline using soft shoreline engineering techniques	Removed concrete shore protection structure in River Canard Park and restored shoreline using soft shoreline engineering techniques, including increasing shoreline crenulaton and excavating a deep water pool; $350,000	2012	Essex Region Conservation Authority, Ontario Ministry of the Environment, and eight other partners	None to date
Riverdance Park, LaSalle, Ontario	Remove derelict marina and restore shoreline for wetland and fish habitat	Stabilized shoreline using soft shoreline engineering techniques along a stretch of the Detroit River shoreline in the Town of LaSalle, including enhancing wetland and fish habitat; $650,000	2009-2010	Environment Canada, Essex Region Conservation Authority, and Town of LaSalle	None to date
Rouge River at Fairway Park, Birmingham, Michigan	Stabilize shoreline using soft engineering techniques, manage woody debris, create a native buffer zone, and remove invasive species	Stabilized two separate 15-meter lengths of stream shoreline, planted a buffer zone of native plants approximately eight meters wide above the bank at both sites, and removed invasive species along the central wooded area between the two plantings; $30,000	2006	Friends of the Rouge and City of Birmingham	Qualitative

Table 9. Cont'd

Rouge River at Ford Field, Dearborn, Michigan	Stabilize eroding stream banks along lower Rouge River and enhance wildlife habitat	Stabilized 274 meters of stream bank using soft engineering techniques (using a live fascine, a brush mattress and a vegetative geogrid), installed rock toe, and planted native species and wildflowers; $108,000	1998–2000	City of Dearborn, Friends of the Rouge, U.S. Environmental Protection Agency, Ford Motor Company, and four other partners	Qualitative
Rouge River at Hines Park, Michigan	Stabilize eroded stream banks and improve fish and wildlife habitat	Stabilized ten severely eroded sections of streambank along 70 meters of shoreline using soft engineering techniques and enhanced 11 hectares of fish and wildlife habitat; total for all ten sites: $780,530; average per site: $78,000	2003–2004	Wayne County Department of Environment and Department of Public Services Parks Division	Qualitative
Rouge River at Shiawassee Park, Farmington, Michigan	Stabilize the riverbank with soft engineering techniques and woody debris, create an adjacent buffer zone of native plants, and enhance aquatic habitat	23 meters of the riverbank was stabilized by grading back the bank and burying bundles of dormant shrubs (live fascines) in the bank, planted a buffer zone of native plants approximately eight meters wide above the bank at both sites and removed invasive species along the central wooded area between the two plantings; $10,000	2004	City of Farmington, City of Farmington Hills, Michigan Department of Environmental Quality, Friends of the Rouge, and seven other partners	Qualitative
Rouge River Oxbow at Greenfield Village, Dearborn, Michigan	Restore fish and wildlife habitat, including wetlands	Restored 671 meters of oxbow shoreline, 1.2 hectares of wetlands and four hectares of uplands; $2 million	Oxbow construction: 2002; fish stocking: 2003	Wayne County, The Henry Ford, Clean Michigan Initiative, and six other partners	Quantitative

Table 9. Cont'd

Solutia Plant, Trenton, Michigan	Stabilize shoreline and enhance habitat	Stabilized berm walls on two existing ponds located on the Detroit River using a variety of limestone riprap to enhance shoreline habitat (in lieu of concrete breakwalls or steel sheet piling); $50,000	2000	Solutia Chemical Company	None
South Fishing Pier, Belle Isle, Detroit, Michigan	Enhance fish habitat at South Fishing Pier	Create one ha of nursery habitat downstream of Belle Isle Sturgeon reef and enhance habitat at South Fishing Pier; $500,000	2012-2013	Detroit Parks and Recreation, Detroit River Remedial Action Plan, Friends of the Detroit River, and others	None
St. Rose Beach Park, Windsor, Ontario	Stabilize shoreline and enhance wildlife habitat	Reconstructed shallow beach area, replaced concrete retaining wall with a rock riprap shore, and added fish habitat features; $196,000	2000–2001	City of Windsor and Essex Region Conservation Authority	Quantitative
Stream crossing at Humbug Marsh Unit, Trenton, Michigan	Build a stream crossing to connect the Refuge Gateway with Humbug Marsh Unit, including the use of vegetated gabion baskets as wing walls to ensure stability and enhance stream bank habitat	Installed a four-meter aluminum box culvert that included 4×3 meter wing walls and planted seedlings of Red Osier Dogwood and black willow to further increase stability and enhance habitat; $30,000	2008	Navy Seabees, Mid-American Group, NTH Consultants, Logs to Lumber & Beyond Inc., and DTE Energy	None

Table 9. Cont'd

Street-End Parks, Trenton, Michigan	Construct three street-end parks and enhance fish habitat to improve fishing opportunities	Created three pocket parks, stabilized shoreline and rehabilitated habitat in the Detroit River; $816,000	2001–2002	City of Trenton, Clean Michigan Initiative, Michigan Natural Resources Trust Fund, and Michigan Coastal Zone Management Program	None
U.S. Steel Shoreline near 203-cm Rolling Mill, River Rouge, Michigan	Restore shoreline using soft shoreline engineering techniques and enhance adjacent riparian habitats	Restore 335 m of shoreline, 0.7 ha of emergent wetlands, and 304 ha of rock shoal; create 0.4 ha of fish spawning habitat and restore 1.9 ha of upland habitat adjacent to the shoreline; $1.2 million	2012-2013	Detroit River Remedial Action Plan, Friends of the Detroit River, USS, and others	None
U. S. Steel Shoreline West of Belanger Park, River Rouge, Michigan	Restore shoreline using soft shoreline engineering techniques and enhance fish and wildlife habitat	Restored 610 meters of Detroit River shoreline; created wetlands that provide spawning and fingerling habitat, and created an upland buffer area to provide water quality protection; $211,000	2004–2005	U. S. Steel, Nativescape, and U.S. Fish and Wildlife Service	Qualitative
Valley Woods Nature Preserve, Southfield, Michigan	Improve green infrastructure for treatment of urban storm water, improve wildlife habitat, and enhance green space	Restore 2 ha of wetlands to increase the capacity to capture and treat storm water and to improve habitat; $300,000	2010-2012	Alliance of Rouge Communities, City of Southfield, Wayne County, The Erb Foundation, Friends of the Rouge, and U.S. Environmental Protection Agency	None to date, two years of post-project monitoring funded

Table 9. Cont'd

William G. Milliken State Park, Detroit, Michigan	Demonstrate innovative storm water management and aquatic habitat rehabilitation	Constructed an innovative storm water retention basin that treated runoff from adjacent neighborhood and rehabilitated shoreline habitat using soft engineering techniques; $1 million	2008–2009	Michigan Department of Natural Resources, Detroit Riverfront Conservancy and Michigan Department of Environmental Quality	None
Windsor Riverfront (Langlois Avenue), Windsor, Ontario	Stabilize shoreline and enhance fish habitat	Created a sloping rock revetment, sloping rock beach and submerged shoal features; planted native species; $800,000	2001	City of Windsor -Department of Parks and Recreation, Essex Region Conservation Authority and Detroit River Canadian Cleanup Committee	Qualitative
Windsor Riverfront – Legacy Park (near Caron Avenue) Windsor, Ontario	Stabilize shoreline and enhance fish habitat	Created a sloping rock revetment, cobble and sand beach, sheltering structures and submerged shoal features; planted native species; $3.4 million	2007	Essex Region Conservation Authority, City of Windsor-Department of Parks and Recreation, and Detroit River Canadian Cleanup Committee	Qualitative
Zug Island, at the confluence of the Rouge and Detroit Rivers, River Rouge, Michigan	Stabilize shoreline of Zug Island and enhance aquatic habitat	Placed recycled bricks from steel plant in front of existing concrete shoreline to create habitat for aquatic life and to serve as a berm to further protect the shoreline from erosion; $10,000	2000	U .S. Steel Corporation	None

partners, including agencies and organizations interested in enhancing or restoring fish and wildlife populations, and their requisite habitats. It was important to involve scientists and resource managers during the initial project planning to broaden the scope of shoreline restoration to include ecological goals. In the case of the Lake Muscoday shoreline restoration projects on Belle Isle in Detroit, projects leaders involved the Natural Resources Conservation Service's Soil Bioengineering Team and University of Michigan faculty to help plan and carry out these restorations. All projects were undertaken as demonstration projects that helped attract partners that wanted to learn new techniques and to help demonstrate community benefits. Further, broadening the project partners helped bring in new funders that leveraged existing funding.

The more than US $24 million cost of the 53 soft shoreline engineering projects identified in Table 9 underscores the need for adequate assessment and pre- and post-project monitoring of effectiveness. One way of accomplishing this is to incorporate pre- and post-project monitoring of effectiveness into all federal, state, and provincial permits for habitat modification (Hartig et al., 2010). Again, the soft shoreline engineering projects identified in Table 9 were frequently undertaken with many partner organizations. One additional suggestion is to work through these partnerships to establish a pre- and post-project monitoring protocol to measure project effectiveness (Hartig et al., 2010). This could be laid out in a Memorandum of Understanding or a non-binding partnership agreement. Greater emphasis should be placed on attracting university students to get involved through independent studies, directed studies, master's theses, practica, and class projects, and also on involving nongovernmental organizations and conservation clubs to use "citizen science" to monitor ecological effectiveness (Hartig et al., 2010).

Habitat restoration, to a close approximation of its original state or to a desired future state, is experiencing a groundswell of support throughout Canada and the United States (Hartig et al., 2011). The number of river shoreline, stream bank, and lakefront restoration projects increases annually. However, far too many of these restoration and enhancement projects have been started without clear definition of restoration goals and quantitative targets for success (Covington et al., 1999). Based on the survey of 53 soft shoreline engineering projects, habitat restoration targets and measurable endpoints were lacking. Therefore, greater emphasis should be placed on ensuring a clear, measurable, ecological definition of project success that includes quantifying habitat/ecological targets and objectives that can be used to both evaluate and select appropriate habitat restoration and rehabilitation techniques, and to measure project success (Hartig et al., 2011).

Most of the soft shoreline engineering projects were undertaken opportunistically through a variety of management tools to enhance/improve riparian or aquatic habitat, including: erosion protection; protection of roads; nonpoint source control; Supplemental Environmental Projects (i.e. a regulatory tool that implements an environmental improvement project instead of paying fines and penalties to a general fund); contaminated sediment remediation; improvement of waterfront parks; enhancement of private developments; dam removal; "green infrastructure" projects; and greenway trail projects. However, there is also a need to move beyond opportunistic habitat rehabilitation and enhancement, and achieve scientifically-defensible, ecosystem-based management. This will require greater identification, quantification, and understanding of essential habitats as a prerequisite to successful management of target species and assemblages. Baird (1996) has shown that lack of scientific understanding and institutional problems are major impediments to scientifically-defensible management of coastal habitats. Further, Baird (1996) recognized the enormous management challenge of shifting from managing species/assemblages to managing habitats to support species/assemblages, particularly in an environment of limited resources for research and management infrastructure.

Actions to rehabilitate and enhance degraded habitats should be based on the understanding of causes and predicted results (Hartig et al., 2011). Adequate assessment, research, and monitoring are essential to define problems, establish cause-and-effect relationships, evaluate habitat rehabilitation and enhancement options, select preferred rehabilitation and enhancement techniques, and document effectiveness. Such assessment, research, and monitoring are the foundation of ecosystem-based management, and, in the end, have often proven to save money for both the public and private sectors (Zarull, 1994).

Only six of the 53 soft shoreline engineering projects surveyed had quantitative assessment of ecological effectiveness (Table 9). Based on a review of these six projects, four had quantitative monitoring that was undertaken opportunistically with no pre-designed plan for monitoring ecological results relative to project goals and objectives (Hartig et al., 2011). Two of these six projects had quantitative monitoring performed to track ecological results relative to project goals and objectives as part of the pre-designed project plan. The monitoring performed at all six projects with quantitative assessment of ecological effectiveness was undertaken for only one or two years. Experience has shown that there is a need to perform long-term monitoring to fully document ecological results (Hartig et al., 2011). Opportunistic monitoring is better than none, but greater emphasis must be placed on strategic monitoring based proper assessment, quantitative target setting, and rigorous post-project assessment of effectiveness as part of an adaptive management strategy. Such post-project monitoring should remain in place for some time as recovery may be slow and adjustments to management actions may be necessary, when implementing an adaptive management strategy (Hartig et al., 2010). Further, there is a need for stronger coupling of habitat modification and research and monitoring. It would be prudent to treat habitat modification projects as experiments that promote learning, where hypotheses are developed and tested using scientific rigor.

Ecological Benefits

The zone referred to as the land-water interface is typically a diverse and productive area. The life cycles of many aquatic and terrestrial organisms are intrinsically entwined with the unique physical and chemical conditions, and interrelationships, that occur only in the vicinity of the water's edge (North Shore of Lake Superior Remedial Action Plans, 1998). In urban areas like Detroit and Windsor metropolitan areas, it is in this land-water interface zone where an integrated approach to redevelopment and restoration can reap numerous ecological benefits.

All people, whether scientists or nonscientists, have some understanding of habitat. It is a place like a river, pond, woods, wetland, or meadow where environmental conditions are right for life, growth, and reproduction of the plants and animals that dwell there (Hartig et al., 2010). Put another way, habitat is a place or set of places where a single organism, a population, or an assemblage of species can find the physical, chemical, and biological features needed for life, including suitable water quality, passage routes, spawning grounds, feeding and resting sites, and shelter from predators and adverse conditions (Federal Interagency Stream Restoration Working Group, 1998). From a resource management perspective, habitat is the physical substrate that supports a biological community of organisms. In the aquatic environment, habitat is commonly depicted as three-dimensional, including both the physical substrate and overlying waters. For all life, habitat is home.

From an ecological perspective, soft shoreline engineering provides much needed aquatic habitat that provides many ecosystem services. Ecosystem services are the processes by which the environment produces resources that we often take for granted, such as clean water, timber, and habitat for fisheries, flood and erosion control, and others. Simply put, habitat conservation and restoration are undertaken to help make sure the benefits of our natural resources—or ecosystem services—are available for healthy coastal communities and future generations.

Soft shoreline engineering is particularly important in channelized rivers like the Detroit River because of the amount of shoreline hardened with concrete breakwaters and steel sheet piling. Soft engineered shorelines can provide spawning and nursery habitat for many fishes, and are critically important to larval fish as they provide shelter, resting areas, food, and a chance to grow a little bigger and stronger on a larval fish's trip downstream to Lake Erie. They can also provide other wetland functions like serving as living filters of nutrients and sediments from land runoff, stabilizing shorelines and minimizing erosion, assisting with groundwater recharge, improving water quality, and more.

Social Benefits

While soft shoreline engineering is important to improving aquatic habitat, it is also important from a social perspective because it can help reconnect people with the natural world. Soft shoreline engineering is increasingly becoming a vital element in making places special or unique and in helping create a much sought-after "sense of place" (i.e. a characteristic held by people that makes a place special or unique; that fosters a sense of authentic human attachment and belonging) on waterfronts in major metropolitan areas. That, in turn, helps contribute to a sustainable community and helps develop more support for restoration and conservation programs.

Indeed, waterfronts are magical places where the water meets the land and people can reconnect with their watershed. Experience has shown that reconnecting people to the river by creating waterfront vistas, reintroducing watershed residents to river history, geography, and ecology, establishing unique conservation places linked by greenway trails and blueways (i.e. canoe and kayak trails), promoting ecotourism, and championing green waterfront developments also help build a political base for a sustainable community.

Economic Benefits

Finally, we cannot lose sight of the economic benefits. Environment Canada has performed economic studies of greenways, natural areas, and parks on the Canadian side of the Detroit River in Windsor, Ontario. These studies show that the closer houses are to greenways, natural areas, parks, and gathering places for wildlife and people, the higher the property values.

In recent years there has been growing interest in estimating and measuring the benefits of restoration efforts. For example, the economic benefits of cleanup of polluted rivers and harbors in the Great Lakes are considerable, but not widely acknowledged. During the period 1990-2004, Canada's Great Lakes Sustainability Fund and its Great Lakes Cleanup Fund spent over $82.8 million (Canadian) on cleanup of Canadian Great Lakes Areas of Concern (i.e. polluted rivers, harbors, and embayments)(Environment Canada, 2004). During the period 1990-2002 alone, these expenditures on federal cleanup activities leveraged another $189.1 million from other partners (Environment Canada, 2004). The money spent in 1990-2004 by the Government of Canada and its partners resulted in the following economic benefits (Environment Canada, 2004):

- An increase in total employment of 7,076 person years (full-time equivalent jobs);
- A $174 million increase in wages and salaries in Ontario;
- A $257 million increase in Gross Provincial Product;
- An $81 million increase in federal, provincial, and local tax revenues;

- Substantial tax savings (four Areas of Concern saved taxpayers of $70 million in capital costs by optimizing sewage treatment plants);
- Increased recreational value;
- Increased property values; and
- Other benefits.

In the U.S., Austin et al., (2007) estimated that a $26 billion investment in cleanup of the Great Lakes through the Great Lakes Regional Collaboration would result in $50 billion in long-term economic benefits. Such economic benefits data provide compelling rationale for investing in restoration projects and programs. Further, such "return on investment" information can be a catalyst for accelerating restoration. There is no doubt that if these soft shoreline engineering projects are treated like experiments and post-project data on ecological, economic, and societal benefits collected, this information could be used as the rationale to further the practice of soft shoreline engineering in other areas. Every effort should be made to communicate and disseminate project benefits and successes broadly through public events and the media, to help further the practice of soft shoreline engineering in other areas.

Clearly, these waterfront green spaces and vistas with soft engineered shorelines also bring benefits through additional recreational spending and increased commercial activity. The economic importance of this was highlighted by the Outdoor Industry Foundation (2006) that quantified that outdoor recreation contributes $730 billion annually to the U.S. economy and supports nearly 6.5 million jobs across the United States.

Beyond the Boundaries

There is another benefit of the 53 soft shoreline engineering projects, particularly from a refuge perspective. The U.S. Fish and Wildlife Service has been advocating working with partners "beyond the boundaries" of refuges to conserve entire landscapes. There is no doubt that what goes on outside the designated border of refuges can have a profound effect on the wildlife, water, and the other resources it was established to protect. Historically, most of the conservation work performed by federal employees was undertaken on federal lands. Today, however, more emphasis is being placed on working with partners on conservation projects on private and non-federal lands outside refuge boundaries in order to protect wildlife corridors consistent with landscape-scale conservation and ecosystem-based management.

The National Wildlife Refuge Association (2005) has reported that refuges are increasingly isolated and squeezed by sprawl, improper development, and some types of agriculture, and has called for more to be done to conserve land and water around refuges. Of the 53 soft shoreline engineering projects completed since 1998, 52 have been on non-refuge lands with other partners. Clearly, the Detroit River International Wildlife Refuge has been a leader of and catalyst for conservation efforts "beyond its boundaries." These conservation projects undertaken "beyond the boundaries" will certainly benefit fish and wildlife, and their requisite habitats, on landscape and ecosystem scales. And it must be remembered that such "beyond the boundaries" conservation projects will benefit both the present generation and others to follow.

Concluding Remarks

The 53 soft shoreline engineering projects completed in the watershed of the Detroit River and western Lake Erie were undertaken for a variety of reasons and employed a number of different approaches or management techniques to enhance/improve riparian or aquatic habitat. All provide "teachable moments" for the value and benefits of habitat enhancement and restoration. However, the value of such projects, both from an ecological and a socio-economic point of view, could be improved by addressing the following key lessons learned:

- Involve habitat experts up front in the design phase of waterfront planning;
- Establish broad-based objectives for shoreline engineering with quantitative targets for project success;
- Ensure sound multidisciplinary technical support throughout the project (e.g. the Natural Resources Conservation Service's Soil Bioengineering Team);
- Start with demonstration projects and attract many partners to leverage resources;

- Treat habitat modification projects as experiments that promote learning, where hypotheses are developed and tested using scientific rigor;
- Involve citizen scientists, volunteers, university students, and/or researchers in monitoring, and obtain commitments for post-project monitoring of effectiveness up front in project planning;
- Measure benefits and communicate successes; and
- Promote education and outreach, including public events that showcase results and communicate benefits.

Although it is encouraging to see 53 soft shoreline engineering projects completed since 1998, it is hoped that the use of soft shoreline engineering will continue to progress and reach the point where it becomes a first option considered in shoreline rehabilitation and urban waterfront redevelopment projects. Regulatory agencies like U.S. Army Corps of Engineers, Michigan Department of Environmental Quality, U.S. Environmental Protection Agency, Public Works Canada, Ontario Ministry of the Environment, and Environment Canada need to recognize and encourage soft shoreline engineering as a best management practice and include it in standard operating manuals for shoreline rehabilitation and urban waterfront redevelopment projects.

It has been stated that a rose that grows surrounded by concrete and steel is more remarkable than one that grows in a horticulturist's garden. If that is the case, then the Detroit River's soft shoreline engineering sites should be celebrated, valued, cherished, and emulated because of the many benefits. Much like the effort to recreate front porches on houses in cities to encourage a sense of community, soft engineered shorelines along waterfronts in urban areas can help recreate gathering places for both wildlife and people.

Literature Cited

Austin, J.C., Anderson, S., Courant, P.N., Litan, R.E., 2007. *Healthy Waters, Strong Economy: The Benefits of Restoring the Great Lakes Ecosystem.* Brookings Institution,Washington,D.C..http://www.brookings.edu/~/media/Files/rc/reports/2007/0904gleiecosystem_austin/0904gleiecosystem_austin.pdf (January 2010).

Baird, R.C., 1996. Toward new paradigms in coastal resource management: linkages and institutional effectiveness. Estuaries 19(2a), 320-335.

Canada Ontario Agreement Respecting Great Lakes Water Quality and Michigan Department of Natural Resources, 1991. Detroit River Remedial Action Plan – Stage 1. Lansing, Michigan and Sarnia, Ontario, Canada.

Caulk, A.D., Gannon, J.E., Shaw, J.R., Hartig, J.H., 2000. Best management practices for soft engineering of shorelines. Greater Detroit American Heritage River Initiative, Detroit, Michigan, USA.

Covington, W.W, Niering, W.A., Starkey, E.,Walker, J., 1999. Ecosystem restoration and management: scientific principles and concepts, p. 599-616. In: Proceedings of the Workshop toward a Scientific and Social Framework for Ecologically Based Stewardship of Federal Lands and Waters. 1995 December 4-14. Tucson, Arizona.

Environment Canada, 2004.The economic benefits of the Great Lakes Sustainability Fund. Toronto, Ontario, Canada.

Federal Interagency Stream Restoration Working Group, 1998. Stream Corridor Restoration: Principles, Processes, and Practices. Washington, D.C.

Fuller, J.A., Gerke, B.E., 2005. *Distribution of shore protection structures and their erosion effectiveness and biological compatibility*. Ohio Department of Natural Resources, Sandusky, Ohio, USA.

Hartig, J.H., Kerr, J.K., Breederland, M., 2001.Promoting soft engineering along Detroit River shorelines. *Land and Water: The Magazine of Natural Resource Management and Restoration* 45(6), 24–27.

Hartig, J.H., Zarull, M.A., Ciborowski, J.J.H., Gannon, J.E., Wilke, E., Norwood, G., Vincent, A., 2007. *State of the Strait: Status and Trends of Key Indicators*. Great Lake Institute for Environmental Research, Occasional Publication No. 5, University of Windsor, Ontario, Canada.

Hartig, J.H., Zarull, M.A., Corkum, L.D., Green, N., Ellison, R., Cook, A., Norwood, G., Green, E., (Eds.), 2010. *State of the Strait: Ecological Benefits of Habitat Modification.* Great Lakes Institute for Environmental Research, Occasional Publication No. 6, University of Windsor, Ontario, Canada, ISSN 1715-3980.

Hartig, J.H., Zarull, M.A., Cook, A., 2011. Soft shoreline engineering survey of ecological effectiveness. Ecological Engineering 37, 1231-1238.

Livchak, C., Mackey, S.D., 2007. Lake Erie shoreline hardening in Lucas and Ottawa counties, Ohio. In: J.H. Hartig, M.A. Zarull, J.J.H. Ciborowski, J.E.Gannon, E.Wilke, G. Norwood, A. Vincent (Eds.), *State of the Strait: Status and Trends of Key Indicators*, pp. 85-90. Great Lakes Institute for Environmental Research, Occasional Publ. No. 5, University of Windsor, Ontario, Canada.

Manny, B.A., 2003. Setting priorities for conserving and rehabilitating DetroitRiver habitats. In: J.H. Hartig (Ed.), *Honoring Our Detroit River: Caring for Our Home*, pp. 79–90. Cranbrook Institute of Science, Bloomfield Hills, Michigan, USA.

Manny, B.A., 2007. Detroit River coastal wetlands. In: J.H. Hartig, M.A. Zarull, J.J.H. Ciborowski, J.E.Gannon, E.Wilke, G. Norwood, A. Vincent (Eds.), *State of the Strait: Status and Trends of Key Indicators*, pp. 172-176. Great Lakes Institute for Environmental Research, Occasional Publ. No. 5, University of Windsor, Ontario, Canada.

National Wildlife Refuge System, 2005. State of the System: An Annual Report on the Threats to the National Wildlife Refuge System.Washington, D.C.

North Shore of Lake Superior Remedial Action Plans, 1998. Achieving Integrated Habitat Enhancement Objectives: A Technical Manual. Thunder Bay, Ontario, Canada.

Outdoor Industry Foundation, 2006. *The Active Outdoor Recreation Economy*. Boulder, Colorado, USA.

Reston Association, 1996. Shoreline stabilization guidelines. Reston, Virginia.

Schneider, J., Heaton, D., Leadlay, H., 2009. Extent of hardened shoreline. State of the Lakes Ecosystem Conference, U.S. Environmental Protection Agency, Chicago, Illinois and Environment Canada, Toronto, Ontario, Canada.

U.S. Fish and Wildlife Service, 2005. Comprehensive Conservation Plan and Environmental Assessment for the Detroit River International Wildlife Refuge. Grosse Ile, Michigan, U.S.A.

Zarull, M.A., 1994. Research: the key to Great Lakes rehabilitation. Journal of Great Lakes Research. 20, 331-332.

CHAPTER 6

Transformation of an Industrial Brownfield into a Gateway to the International Wildlife Refuge

The first time I entered the site I had an eerie feeling. Barbed wire fences encircled the entire site, including the shoreline of the Detroit River. It was flat as a pancake, with few trees. The few trees that were present were along the fence lines, as if standing guard and warning people to stay out of this industrial brownfield. A brownfield is an abandoned or underused industrial or commercial site whose future use is affected by historical residual contamination. Between the barbed wire and my perception of the perimeter trees standing guard, I clearly felt as if I was trespassing on property where I didn't belong.

A concrete road divided much of the site, but even the concrete was showing its age with many cracks and crevices similar to the wrinkled face of an old man. Every so often I would see some discarded evidence of its former industrial glory, pieces of steel, sheet metal, coiled rubber coated cables, and broken pallets. In the northeast corner of the site lay the old concrete pad, upon which part of the paint and brake plant manufacturing facility sat. A few mulberry bushes were actually growing through the cracks in the concrete pad, as if saying let me out or set me free. The site must have been something in its industrial heyday.

Industrial History

In 1946, the Chrysler Corporation applied to the City of Trenton, Michigan for a permit to construct its first building on 17.8 ha of vacant land along the Trenton Channel of the Detroit River (Table 10). Production of adhesives, sealers, and paints began in 1947 under the brand name of "Cycleweld" (Driscoll and Elliott, 1990). Additional buildings were constructed on site from 1952 through 1973. Brake linings were first produced in the chemical building as a pilot program in 1958. Because of the success of this pilot program and an increased demand for these products, the brake lining division expanded into a new Friction Products Building in 1964 (Driscoll and Elliott, 1990). This operation further expanded in 1968.

Table 10. Site history of Wayne County's Refuge Gateway in Trenton, Michigan.

Date	Activity
1946-1990	• Chrysler operated this facility as an automotive component manufacturing plant, including production of brakes, associated adhesives, oils, and sealers
1990	• Chrysler facility closed and remediated to industrial standards
1994	• Consent Decree signed, with restrictive covenants
2001	• Detroit River International Wildlife Refuge Establishment Act signed into law
2002	• Chrysler property purchased by Wayne County with National Oceanic and Atmospheric Administration funding to become the Refuge Gateway for the Detroit River International Wildlife Refuge • Consent Decree modified to allow for the construction of a Visitor Center on site for the Detroit River International Wildlife Refuge
2004	• Wayne County completes site Master Plan for Refuge Gateway, with input from U.S. Fish and Wildlife Service and other partners
2005	• Comprehensive Conservation Plan completed and approved by U.S. Fish and Wildlife Service, identifying Wayne County's Refuge Gateway as the proposed site of a future Refuge headquarters and visitor center
2006	• Schematic Plan for Refuge Gateway completed, providing more detailed information for restoration efforts and visitor center design, and integrating the Refuge Gateway property with Humbug Marsh Unit of the U.S. Fish and Wildlife Service
2006-2007	• Kresge Foundation grant for planning for green site features at Refuge Gateway and LEED-certified visitor center
2007	• State of Michigan unveils historical marker at Refuge Gateway, identifying it as the first location on the Michigan Conservation Trail that commemorates and promotes knowledge of conservation history in Michigan
2008	• Trails, environmental education shelter, and wetland boardwalk constructed in Humbug Marsh Unit and pedestrian stream crossing completed, linking Humbug Marsh Unit and Refuge Gateway
2009	• Daylighting Monguagon Creek completed and capping and final grade achieved on 40% of the Refuge Gateway • First loop of access road completed at the Refuge Gateway
2010	• Perimeter greenway trail completed at Humbug Marsh, linking Refuge's Lake Erie Metropark Unit with Humbug Marsh Unit and the Refuge Gateway • Entry garden completed at Refuge Gateway by volunteers from Ford Motor Company, Grosse Ile Garden Club, and International Wildlife Refuge Alliance

Table 10. Cont'd

2011	• Shoreline and riparian buffer habitat restoration completed • Final grade achieved on another 30% of the Refuge Gateway • Construction of a shoreline access road to provide accessibility for the future boat dock and fishing pier • Creation of a kayak landing
2012	• All brownfield cleanup work completed • Over 300 native trees planted, most with over 136-kg (300-pound) rootballs
2013	• One ha of land transferred to the U.S. Fish and Wildlife Service to construct the visitor center • Groundbreaking ceremony held for visitor center • 150 additional native trees planted

The Chrysler friction products plant produced both asbestos and non-asbestos brake linings for Chrysler automobiles. Asbestos and other components (e.g. zinc powder, lead powder, cellulose filler, steel fibers, graphite, and phenolic resin), which varied depending upon the specific product formulation, were weighed and transported in dumpsters to a mixing area (Driscoll and Elliott, 1990). In this area, the materials were mixed in ribbon blenders and returned to dumpsters for temporary storage, followed by transport as a dry mix formulation.

Depending upon the type of friction product manufactured, the formulation may have been extruded or pressed into the shape of the brake component (Driscoll and Elliott, 1990). The formed brake pads or linings were transported to cutting ovens and/or additional presses. The cured brake products underwent a variety of processing operations, including sawing, grinding, and drilling.

Over the history of this automotive parts facility, the Chrysler Corporation manufactured brake pad adhesives for automobiles, blended oils, paints, sealers, powdered metal parts, asbestos brake pads, and phenolic brake pistons (Figure 17). Historical aerial photographs show that fill was placed on the eastern portion of the property from the 1940s to 1967. This literally filled approximately 4.5 ha of wetlands and extended the upland portion of the property eastward into the Detroit River approximately 0.25 km.

Figure 17. Aerial photograph of the Chrysler Manufacturing Facility in Trenton, Michigan, 1967 (photo credit: City of Trenton).

After 44 years of operation, this Chrysler Manufacturing Plant was deactivated in 1990 and underwent removal of all above ground structures (Table 10). In 1994, Chrysler Corporation and the Michigan Department of Natural Resources entered into a Consent Decree. The Consent Decree is a legally binding document describing Chrysler Corporation's remedial responsibilities, the extent of continuing liability for this site, and long-term due care obligations (Stell, 1994). On-site remedial activities included removal and isolation of both inorganic (i.e. asbestos, arsenic, barium, cadmium, lead, cyanide and thallium) and organic (i.e. benzene, chlorobenzene, methylene chloride, toluene, vinyl chloride and xylene) contaminants.

Essentially, the site was decommissioned and cleaned up at a cost of approximately $12 million to meet State of Michigan standards for industrial and commercial use. Pursuant to the Consent Decree, restrictive covenants were placed on five areas of the site totaling approximately 6.2 ha (15 acres) (Stell, 1994). These restrictive covenants limit future activities and use of the five areas due to subsurface contamination. For example, there can be no digging in the capped areas that might release contaminants or exasperate environmental problems. Restrictive covenants remain in effect in the transfer of the property to any future owner. Therefore, Wayne County is required to follow these restrictions determined in the

Consent Decree. Specific restrictions placed on these five areas of the property included:

- Any soils removed from the site must be tested for contaminants;
- On-site groundwater may not be used as drinking water;
- Uses of the property must be restricted to industrial uses consistent with the remediation performed (any other uses may require reevaluation and the State of Michigan);
- The State of Michigan must have access to determine compliance, including the rights to take samples, inspect records, and inspect remedial actions; and
- In five restricted areas, future owners must restrict activities that might interfere with a response activity, operation and maintenance, monitoring, or other measures necessary to assure the effectiveness and integrity of the remedial actions.

In addition, there is a due care plan for the site that helps ensure that all restrictions and measures in the Consent Decree are followed in perpetuity, and that any future uses do not exasperate residual environmental contamination.

Preservation of Humbug Marsh

Located adjacent to Trenton Chrysler Plant was the 166-ha Humbug Marsh – a coastal marsh, a barrier island, and uplands that had a long hunting tradition and was even farmed at one time. In the mid-1990s, a developer purchased Humbug Marsh with the intent of building a subdivision, a marina, a golf course, a bridge to Humbug Island to construct homes, and more. To proceed with the proposed development required several permits and requisite public hearings. The first public hearing was held at Gibraltar Carlson High School in September 1998 and attracted nearly 1,000 people (U.S. Fish and Wildlife Service and International Wildlife Refuge Alliance, 2005). People from all over Michigan attended, creating traffic jams and causing the fire marshal to lock the doors of the school to prevent a larger crowd than allowed by fire safety regulations. The issue at hand was residential development of the last kilometer of natural shoreline on the U.S. mainland of the Detroit River – Humbug Marsh.

The vast majority of these citizens strongly opposed the development and was in favor of preserving the rich and diverse coastline that was part of their home and heritage. Indeed, citizens and grassroots organizations banded together for nearly ten years in a campaign to preserve Humbug Marsh. This tremendous public support was a key catalyst in establishing the DRIWR. The people spoke out in opposition, the permits were not issued, and eventually Humbug Marsh was purchased in 2004 for the DRIWR. Humbug Marsh now is part of the National Wildlife Refuge System and is protected for wildlife and for people to enjoy through wildlife-compatible public uses. Today, it stands as a site of great determination and pride by those in the region.

Following the preservation of Humbug Marsh in 2004, the Michigan Department of Environmental Quality and the U.S. Fish and Wildlife Service spent nearly three years compiling scientific data on Humbug Marsh that were used as the rationale for obtaining a "Wetland of International Importance" designation under the international Ramsar Convention. In 2010 Humbug Marsh was designated as a Ramsar Convention "Wetland of International Importance." There are over 2,100 such Ramsar designations worldwide, 36 in the United States, and only one in Michigan – Humbug Marsh. The Ramsar Convention is an international treaty that was signed in Ramsar, Iran in 1971 that provides a framework for voluntary international protection of wetlands. Countries signing the treaty must demonstrate their commitment to the conservation and wise use of wetlands as a contribution toward sustainable development throughout the world. Humbug Marsh is considered an internationally important wetland because of its ecological importance in the Detroit River corridor and the Great Lakes Basin Ecosystem (Table 11). It serves as vital habitat for 51 species of fish, 90 species of plants, 154 species of birds, seven species of reptiles and amphibians, and 37 species of dragonflies and damselflies.

Transformation of the Refuge Gateway

Following the establishment of the DRIWR in 2001 and during the public campaign to preserve Humbug Marsh, the 17.8-ha, former Chrysler property in Trenton was purchased to become the Refuge Gateway (Table 10). The Refuge Gateway was given its name to become the gateway to the international wildlife refuge and to become the future home of the refuge's visitor center (Hartig et al., 2012).

Table 11. Ramsar criteria and rationale for designating Humbug Marsh a "Wetland of International Importance."

Ramsar Criterion	**Rationale**
Importance to Threatened, Endangered and Vulnerable Species, and Ecological Communities	The wetlands within Humbug Marsh are classified as Great Lakes marsh, a natural community that has been ranked as a globally imperiled community by the Michigan Natural Features Inventory. As the shorelines of the Great Lakes were developed for industrial, commercial, residential, and recreational use, the marsh habitat essential for many Great Lakes species rapidly declined.
Importance for Maintaining Biological Diversity	Because Humbug Marsh represents a significant portion of the last unaltered wetlands in the Detroit River corridor and the last mile of natural shoreline on the river's U.S. mainland, it serves as a vital habitat for a large variety of endemic fish, birds, and plants that are regionally rare and may otherwise be extirpated from the area. Surveys have documented 51 species of native fish, over 90 native plant species, at least a 154 native bird species from 39 different families, more than 25 species of reptiles and amphibians, and 12 species of damselflies and 25 species of dragonflies.
Importance as Habitat for Plants or Animals in Critical Stages of their Lifecycles	Humbug Marsh and the lower Detroit River are located at the intersection of two important migratory bird flyways (i.e. Atlantic and Mississippi Flyways), making it prime stopover habitat during fall and spring migrations. Humbug Marsh is also a critical corridor along the Detroit River for herpetofauna and serves as an important breeding, nesting, and developmental site for a number of amphibian and reptile species.
Importance to Indigenous Fish Biodiversity	The variety of habitats existing within Humbug Marsh allows fish with diverse life histories to thrive. In a fish survey that included Humbug Marsh and surrounding areas, 51 indigenous fish species were identified, representing 15 different families.
Importance as a Food Source, Spawning, Nursery or Migrat-ion Area on which Fish Depend	Each year, over 3 million Walleye (*Stizostedion vitreum*), approximately 10% of the population of Lake Erie, run the Detroit River. Once the Walleye spawn on rocky substrate within the river, larval fish travel to the marsh and use it as an essential nursery habitat. The wetlands provide spawning and nursery areas for Yellow Perch (*Perca flavescens*), Muskellunge (*Esox masquinongy*), Brown Bullhead (*Ameiurus nebulosus*) and many other fishes. In addition, Humbug Marsh serves as one of the only remaining spawning and nursery areas for forage fishes, which rely on the significant cover of emergent and submergent vegetation for their survival.

From the outset, U.S. Congressman John Dingell had served as the project champion, opening doors to key decision-makers, bringing in new partners, and building the capacity for growing and managing the DRIWR.

In 2003, Wayne County initiated a master planning process for the Refuge Gateway that was completed in 2004 (Figure 18). The goal of the master plan was to be a model of sustainable redevelopment by providing a blueprint for all cleanup and restoration work necessary to establish the site as an ecological buffer for Humbug Marsh and to encourage public uses.

Everything people would see and do at the Refuge Gateway would teach them how to live sustainably (Hartig et al., 2010). Further, the site would be integrated with Humbug Marsh in a fashion that would help Wayne County, the U.S. Fish and Wildlife Service, and other partners teach the next generation of conservationists and sustainability entrepreneurs (Hartig et al., 2012).

This mission of teaching the next generation of conservationists and sustainability entrepreneurs was to be accomplished by offering exceptional environmental education, interpretation, and outdoor recreation in an ecosystem that has been recognized for its biodiversity in the North American Waterfowl Management Plan, the United Nations Convention on Biological Diversity, the Western Hemispheric Shorebird Reserve Network, the Biodiversity Investment Area Program of Environment Canada and U.S. Environmental Protection Agency, and the international Ramsar Convention on Wetlands. Further, restoration activities would serve as "teachable moments" in support of the educational mission.

Specific restoration targets for the site included: restoring 6.5 ha of wetlands in a river that has lost 97% of its coastal wetlands to development; restoring 10.1 ha of upland buffer habitat; treating invasive *Phragmites* along 4 km of shoreline; and treatment of invasive plant species in over 20.2 ha of upland habitats in Humbug Marsh (Hartig et al., 2012). Therefore, this project is consistent with ecological theory that calls for protecting habitats of exceptional biodiversity (e.g. Humbug Marsh Unit), expanding their ecological buffers (e.g. restoring fish and wildlife habitat at the adjacent Refuge Gateway), and then linking with other high quality habitats.

DETROIT RIVER INTERNATIONAL WILDLIFE REFUGE GATEWAY

Figure 18. Refuge Gateway Master Plan adopted by Wayne County and partners in 2004 (master plan credit: Hamilton Anderson Associates).

In 2005, the Refuge Gateway was identified in the Refuge's Comprehensive Conservation Plan as the future home of the visitor center (U.S. Fish and Wildlife Service, 2005) and a schematic plan was completed in 2006. Additional site planning for green site features and Leadership in Energy and Environmental Design (LEED) certification was then undertaken in 2005-2007.

As part of the effort to showcase sustainable practices, the Monguagon Creek was unearthed from an underground pipe at the Refuge Gateway and daylighted in 2009 (Hartig et al., 2012). Historically, storm water from a small watershed in Trenton was transported to the Detroit River through a concrete underground pipe. Through this Monguagon Creek Daylighting and Wetland Restoration Project, 2.4 ha of wetlands were created in an area that has lost 97% of its coastal wetlands to anthropogenic development (Figure 19A and B). The project included constructing a storm water treatment basin and an emergent wetland as a two-step process of naturally treating urban storm water before entering the Detroit River. The retention pond accomplishes settling of solids and the emergent wetland promotes nutrient uptake. A pump installed in the wetland also allows for the movement of water from the constructed wetland into the adjacent Monguagon Delta of Humbug Marsh. Moving water into Humbug Marsh assists with decreasing the density of invasive *Phragmites* in the Monguagon Delta, thereby promoting greater native species diversity. Additionally, the delta provides 1.4 ha of established emergent wetland for additional storm water treatment. Other project benefits included wildlife habitat creation and opportunities for hands-on environmental education and natural resource interpretation. Also completed in 2009 were the first access road and some capping, and achievement of final grade, as called for in the master plan, on the western third of the Refuge Gateway property.

Again, through the historical development of the site for industrial purposes during 1940s-1960s, approximately 0.25 km of river bottom and wetlands were filled to support a number of manufacturing operations. At the time of closure of the plant in 1990, the site was relatively flat, with an approximately 3 m drop at a 90° angle to the Detroit River. As noted above, the site master plan called for the restoration of a more natural shoreline at the Refuge Gateway. A coastal wetland and habitat restoration plan was then prepared in 2010 by a team of expert scientists, engineers, designers, and representatives from regulatory agencies, accounting for the site environmental constraints. For example, the Consent Decree stipulated that five sites totaling approximately 6 ha were subject to restrictive covenants where no digging or excavation could occur. Based on a review of the Baseline Environmental Assessment and other background environmental reports on site contamination by the project team, and analytical testing at randomly selected sampling stations on site, the southeastern portion of the property was selected as the most suitable location for excavation and restoration. In an area that has lost an estimated 97% of historical coastal wetland habitat, this shoreline restoration project would represent a regional net gain of 1.2 ha of this threatened habitat.

In 2011, shoreline restoration began with the removal of 30,600 m^3 of fill from the 1.2 ha of land at southeast corner of the property (Hartig et al., 2012). Through the excavation process it was discovered there was a lens of residual contamination (i.e. primarily petroleum residue) at the old river bottom (4.3-4.6 m below existing

Figure 19. Photographs of the Refuge Gateway as a brownfield site prior to daylighting Monguagon Creek (A) and after daylighting (B) (photo credit: U.S. Fish and Wildlife Service).

grade) that was filled in during the 1940s-1960s. It was the consensus of the project term and environmental experts that the residual contamination discovered 4.3-4.6 m below grade at the old river bottom was most probably historical oil pollution dating back primarily to the 1940s-1960s.

To avoid exposing residual contamination and creating a direct hydrologic pathway to the Detroit River, and to comply with regulatory agency cleanup standards and shoreline restoration permits, the shoreline restoration design had to be modified consistent with an adaptive management philosophy. The project team, with concurrence from state and federal regulatory agency staff, agreed to limit excavation to 0.6-1.2 m above the old river bottom (i.e. contaminated lens stratum). This would in essence maintain a cap over the contaminated material. It should be noted that there was no evidence of residual contaminant movement into the Detroit River.

The project team modified the design plans to construct a coastal wetland shelf along the entire Detroit River shoreline and riparian buffer habitat at the southeastern corner of the site. Overland water flow and fluctuating water levels in the Detroit River would maintain seasonal inundation to support a hydrophilic plant community at this southeastern corner. These design changes produced 0.9 ha of coastal wetlands.

No excavated soils were removed from the site. All 30,600 m^3 of soils excavated for restoration of wetland and riparian buffer habitat was deemed environmentally-suitable material for capping and then relocated on site to begin capping of the approximately 6 ha subjected to restrictive covenants. Eventually, at least 0.9-1.8 m of clean fill was placed on all 6 ha subjected to restrictive covenants to meet human health standards and to provide a greater level of protection for wildlife.

Throughout all excavation and restoration work completed during 2008-2012, four additional unforeseen environmental challenges were discovered. These included discovering contaminated sediment while building a stream crossing, unearthing an abandoned underground storage tank while building the first access road, finding seven 208-L drums of solid hazardous waste, and discovering small area of oil-contaminated soil from several rusted barrels. In each case, a proper environmental assessment had to be performed to develop a remedy to meet all state and federal cleanup standards. In the case of the contaminated sediment at the stream crossing, all material had to be removed in lined roll-off boxes and disposed of in a Type 2 Landfill. In the case of the abandoned underground storage tank, analytical work confirmed no residual contamination and the tank was recycled. In the case of the seven 208-L drums of solid hazardous waste and oil-contaminated soil, no scientific evidence of leaching was discovered and all material was removed by environmental response experts from U.S. Environmental Protection Agency and taken to a hazardous waste treatment facility.

The Detroit River shoreline restoration work was completed in 2011, including construction of the second access road and a kayak landing for public use (Hartig et al., 2012). Control of invasive species was initiated in 2008 and will continue in the future to sustain these unique habitats and biodiversity. Also in 2011, funding was secured to complete the remaining capping and upland buffer habitat restoration called for in the master plan. All construction activities for brownfield cleanup and habitat restoration at the Refuge Gateway were completed in 2012, expanding the ecological buffer for Michigan's only "Wetland of International Importance" and laying the ecological foundation for a visitor center. Design of the Leadership in Energy and Environmental Design (LEED)-certified visitor center was completed in 2013 (Figure 20). Construction began in November 2013.

William McDonough, an internationally renowned architect and industrial designer who champions sustainability, has offered planners, developers, and community leaders the following challenge:

> *If design is the signal of human intention, then we must continually ask ourselves—What are our intentions for our children, for the children of all species, for all time? How do we profitably and boldly manifest the best of those intentions?*

This challenge was placed before all stakeholders at the beginning of the project and throughout all work. This became part of the vision that has now turned into the reality of sustainable redevelopment of the Refuge Gateway and preservation of Humbug Marsh into a hub for environmental education and sustainability. Literally over 300 public, private, and nonprofit organizations have made contributions to this effort and the Refuge was singled out as a national leader in public-private partnerships in the 2005 White House Conference on

Cooperative Conservation (Council on Environmental Quality, 2005). Stakeholders have been involved in all aspects of this project from the outset, including planning, design, fundraising, implementation, and stewardship. As a result, it is hoped that a sense of ownership and community pride be felt by all stakeholders, which in turn may help grow a sustainability and stewardship ethic for the region.

Sustainable Public Use

The philosophy of the U.S. Fish and Wildlife Service is to protect wildlife first and then offer wildlife-compatible public uses, when and where possible. Special emphasis is placed on providing quality public uses. As part of the effort to provide quality outdoor educational and recreational experiences, and to teach the next generation of conservationists and sustainability entrepreneurs, 4 km of hiking trails, two wildlife observation decks, a wetland boardwalk, an environmental educational shelter, and unique learning stations were constructed in Humbug Marsh in 2007-2009 (Hartig et al., 2010). Additional trails, an observation decks, a world-class fishing pier and dock for the Great Lakes school ship, and another greenway extension are called for in the Master plan for the Refuge Gateway. These public amenities will be constructed in conjunction with the refuge's visitor center in the future.

Figure 20. Visitor center being constructed at the Refuge Gateway in Trenton, Michigan (rendering credit: LHB Corporation).

The project's focus on sustainability was strongly reflected in material choice. For example, the stream crossing, connecting Humbug Marsh to the Refuge Gateway, was built with used utility poles and ash trees killed by the emerald ash borer, universally accessible trails were constructed with recycled crushed concrete, a wildlife observation deck was constructed with reclaimed wood, and the wetland boardwalk was constructed with 100% recycled plastic wood. As often as possible, local materials and contractors were utilized on the project to reduce emissions from travel and to create meaningful ties to the community.

Concluding Thoughts and Lessons Learned

Standing at the Refuge Gateway is like viewing three different centuries at once:

- To the south is Humbug Marsh – the last kilometer of natural shoreline on the U.S. mainland of the Detroit River that has on its uplands an old growth forest with oak trees over 300 years old that were alive when Cadillac founded Detroit in 1701 (Figure 21A);
- To the north is the Trenton Channel Power Plant that represents the industrial revolution of the 20th Century (Figure 21B); and
- On the site can be seen a 21st Century example of sustainable redevelopment of a 20th Century industrial brownfield into the Refuge Gateway that will be the home of the Refuge's LEED-certified visitor center (Figure 21C).

Many people still view the Refuge Gateway as a paradox of heavy industry and internationally-recognized wildlife refuge. But it is not. It is a strategically-planned destination of choice consistent with the philosophy of Abraham Lincoln who said:

The best way to predict the future is to create it!

This project has resulted in: a net gain of 6.5 ha of wetlands in an area that has lost 97% of its coastal wetlands to development; restoration of 10.1 ha of upland buffer habitat; control of invasive plant species on 20.2 ha of upland habitats; and control of invasive *Phragmites* along 4 km of shoreline. It has also resulted in merging

Figure 21. Three perspectives obtained from the Refuge Gateway in Trenton, Michigan: A) an 18th Century shoreline perspective obtained from the view to the south of Humbug Marsh – the last kilometer of natural shoreline on the U.S. mainland of the Detroit River; B) a 20th Century industrial revolution perspective obtained from the view to the north of the Trenton Channel Power Plant; and C) a 21st Century example of sustainable redevelopment from the transformation of an industrial brownfield into the Refuge Gateway.

the 18-ha Refuge Gateway with the 166-ha Humbug Marsh into one ecological unit and is helping create a truly exceptional outdoor recreational and environmental educational experience.

In this transformational project, it was very important to reach broad-based agreement on a long-term sustainability vision founded on a sense of place (i.e. a characteristic held by people that makes a place special or unique and that fosters a sense of authentic human attachment and belonging). Congressman John Dingell became the project champion who was well recognized and trusted throughout the region, and could open doors, help bring in partners, and build capacity. A core project delivery team coordinated all efforts, including making sure that all regulatory agencies were involved up front in the process to ensure buy-in, support for project, and timely review and approval of permits.

Much effort was expended to recruit and inspire partners to get involved in all aspects of the project (e.g. building a native plant entry garden, planting trees, removing invasive species, etc.), founded on cooperative learning and action. An adaptive management process was used that assesses problems and opportunities, sets priorities, and takes actions in an iterative fashion for continuous improvement. Transparency in the decision-making process was essential. Finally, a high priority was placed on measuring and celebrating progress throughout the process to sustain project momentum, and on inspiring the next generation of conservationists and sustainability entrepreneurs through meaningful service learning (i.e. a strategy that integrates meaningful community service with instruction and reflection to enrich the learning experience, teach civic responsibility, and strengthen communities) and exceptional environmental education.

Literature Cited

Council on Environmental Quality, 2005. Proceedings of the White House Conference on Cooperative Conservation, St. Louis, Missouri. 2005 August 29-31.Washington, DC, USA. http://cooperativeconservation.gov/contact.html (January 2011).

Driscoll, R.J., Elliott, L.J., 1990. Health Hazard Evaluation Report No. 87-126-2019 (Chrysler Chemical Division, Trenton, Michigan). National Institute for Occupational Safety and Health, Cincinnati, Ohio, USA.

Hartig J.H., Robinson, R.S., Zarull, M.A., 2010. Designing a Sustainable Future through Creation of North America's only International Wildlife Refuge. Sustainability 2(9), 3110-3128.

Hartig J.H., Krueger, A., Rice, K., Niswander, S.F., Jenkins, B., Norwood, G., 2012. Transformation of an industrial brownfield into an ecological buffer for Michigan's only Ramsar Wetland of International Importance. Sustainability. 4(5), 1043-1058.

Stell, C., 1994. Administrative Order of Consent, Covenants Not to Sue, and Contribution Protection (Case Number 94-079112-CE). 30th Judicial Circuit Court, Lansing, Michigan, USA.

U.S. Department of the Interior, U.S. Fish and Wildlife Service, U.S. Department of Commerce, U.S. Census Bureau, 2006. National Survey of Fishing, Hunting, and Wildlife-Associated Recreation, Washington, D.C., USA.

U.S. Fish and Wildlife Service, 2005. Comprehensive Conservation Plan and Environmental Assessment for the Detroit River International Wildlife Refuge. Grosse Ile, Michigan, USA. www.fws.gov/midwest/planning/detroitriver/ (September 2011).

U.S. Fish and Wildlife Service and International Wildlife Refuge Alliance, 2005. Building Our International Wildlife Refuge in the Industrial Heartland, Grosse Ile, Michigan, USA.

CHAPTER 7

If You Build It, They Will Come

A nearly two-meter-long female fish, that has remained virtually unchanged since prehistoric times, leaves the deep waters of Lake Huron and heads south in search of her natal spawning grounds in the Detroit River with a determination and intensity similar to that of a snow goose that leaves its summer breeding grounds in the arctic tundra and annually migrates over 4,800 km (3,000 miles) to warmer wintering areas in North America. Using her strongly muscled body and shark-like tail, she slowly and methodically swims toward the very few spawning habitats remaining in the Detroit River. Much of her life throughout the year is spent prowling the cold deep waters of the upper Great Lakes until it is time to spawn. Her species matures very slowly and females do not reproduce until they reach the age of about 25. She is a 30-year old fish in the prime of her reproductive years and, despite her intimidating look and size, is quite docile.

Her species has probably been resident in the Great Lakes since the end of last ice age, about 10,000 years ago. Fishery biologists call her a "living dinosaur" because her species thrived during the Upper Cretaceous period (136 million years ago) when dinosaurs were at the height of their development. She may live for more than150 years, but typically only 40-60 years, and can grow to a length of 2.7 m (9 feet) and a weight of 136 kg (300 pounds). Males of her species live 50-60 years. She has an olive-brown body with a milky-white belly. Her body "skin" is composed largely of bony plates called "scutes" that were probably developed for protection from fish eating "dinosaurs." She is a bottom feeder and has a tube-like mouth and no teeth. She uses her four distinctive fleshy whiskers called "barbells," that hang down in front of her mouth, to smell and taste food in the water. When she finds food while swimming along the bottom, she extends her tube-like mouth and quickly sucks it in. Most of her diet is made up of molluscs, other bottom-dwelling invertebrates, and small fishes. Once sucked in, her prey is strained out by her gill structures and swallowed while the silt or sand, taken in with the food, is flushed out through her gill openings.

She truly is an awesome sight. Native Americans called her species "Ogimaagiigonh" or “King of Fishes” and harvested them in early spring during their spawning season. A barbed, harpoon-like spear was the favored Native American fishing gear for this species during their spawning runs (Cornell, 2003). A speared fish would then be secured to the canoe and both dragged along by the Natives to tire out the fish. Once tired out, the fish was pulled into the canoe and killed by lance or clubbing (Cornell, 2003).

Native Americans also revered them as a culturally significant species. The Wyandot believed that the Spirit named “Oki” lived in fish (Givens-McGowan, 2003). In each fish house in the Wyandot village, there was a medicine man who was skilled at talking to the fish and telling them that they were respected. Respect would be shown by burying the bones, instead of burning them. Medicine men would thank the fish in advance for giving up their lives so that Natives could eat and live (Givens-McGowan, 2003).

In contrast to the reverence bestowed on them by Native Americans, early immigrant fishermen before 1850 viewed this species as a “trash fish” and slaughtered them because they destroyed fishing nets targeting other species. This “trash fish” moniker lasted until their market value as caviar was recognized, which quickly led to the development of an intensive, commercial, fishing industry.

During the heavy fishing years from 1879 to 1900, the commercial catch of this species in the Great Lakes averaged over 1,814 tonnes (4 million pounds)(U.S. Fish and Wildlife Service, 2012). In 1885, a maximum of 4,901 tonnes (8.6 million pounds) were harvested, of which 2,359 tonnes (5.2 million pounds) came from Lake Erie (U.S. Fish and Wildlife Service, 2012).

After the industrial revolution many of their populations were nearly extirpated. In the Great Lakes, humans went on to overfish them and degrade and destroy their habitats, pollute their environment, and block access to their traditional spawning grounds by building dams. Scientists and fishery managers are now working to bring back this native “King of Caviar.”

Today, we call her Lake Sturgeon (*Acipenser fulvescens*) and scientists use her as a barometer of ecosystem health and biodiversity in the Great Lakes Basin Ecosystem. Conservationists and environmentalists refer to the Lake Sturgeon as a charismatic megafauna because of its large size and popular appeal. Even famed poet Henry Wadsworth Longfellow (1855) was inspired by the Lake Sturgeon:

Lay the monster Mishe-Nahme,
Lay the Sturgeon, King of fishes;
Through his gills he breathed the water,
With his fins he fanned and winnowed,
With his tail he swept the sandfloor,
There he lay in all his armor...

Human Impacts

In the late-1600s and early-1700s, French explorers and voyageurs noted that the Detroit River and Lake Erie teemed with Lake Sturgeon and Lake Whitefish (*Coregonus clupeaformis*). Indeed, Sturgeon remained common over a century later as noted by Lanman (1839) in his book *History of Michigan*:

> *These lakes abound also with fish, some of the most delicious kinds. Among these are the Sturgeon, the Mackinaw Trout, the Mosquenonge (*muskellunge*), the white fish, and other of smaller size peculiar to fresh water. The Sturgeon advances up the stream from the lake during the early part of spring to spawn, and are caught there in large quantities by the Indians.*

We know that during the 1800s, Lake Sturgeon were found in all the Great Lakes, but were one of the most abundant fish species in lakes Huron and Erie, including the Detroit River. Commercial fishing log books from this era help provide an important perspective on the abundance of Lake Sturgeon in the Detroit River and Lake Erie. For example, commercial fishing in the waters of Pointe Mouillee off the mouth of the Detroit River in 1877 documented 159, 244, 433, and 214 Lake Sturgeon caught on four separate days in early May, demonstrating its high abundance during this era (Duglas et al., 1877).

Scott and Crossman (1973) noted that the relationship between Lake Sturgeon and humans during the 1800s could best be described as "kaleidoscopic." Before the 1860s, commercial fisherman slaughtered Lake Sturgeon as a nuisance fish because they became entangled in nets, destroying them, and because they sucked up spawn (Bogue, 2000). During this same time period, there were reports that Lake Sturgeon were stacked like firewood and left to dry on the banks of the Detroit River in Amherstburg, Ontario. The mummified bodies of the oily fish were burned to heat the boilers of wood-burning steamboats on the Detroit River (Bein, 2012). Again, it was during the mid-1860s that the economic value of Lake Sturgeon was recognized when their eggs became sought after as caviar and their smoked flesh became craved as a delicacy. This led to the development of an important commercial fishery. Total Lake Sturgeon harvest peaked in the mid-1880s at 4,901 metric tons (8.6 million pounds). Interestingly, in 1890 a "caviar factory" was located in Algonac, Michigan on the St. Clair River (Harkness and Dymond, 1961).

Demand for Lake Sturgeon was so high in the late-1800s that fishery managers even propagated them artificially. A report of the Ohio Game and Fish Commission in 1891 and summarized in Harkness and Dymond (1961) reported that:

> *On May 31st, with the assistance of Mr. Samuel Currie, we got some 6,000,000 Sturgeon eggs and hatched them in boxes in open water about the islands at the mouth of the Detroit River – succeeded in hatching about 5,000,000. Some, we planted at the mouth of the river at various places, and the remainder, about the islands of Put-in-Bay.*

During this late-1800s time period, the corridor between lakes Huron and Erie was one of the most productive waters for Lake Sturgeon in North America. For example, in 1880 lakes Huron and St. Clair produced over 1.8 million kg (four million pounds) of Lake Sturgeon (Hay-Chmielewski and Whelan, 1997). In 1890, Lake Erie produced over 275,000 kg (610,000 pounds) of Lake Sturgeon in Canadian waters alone (Figure 22). During the spawning period in June 1890, upwards of 4,000 adult Lake Sturgeon were caught in Lake St. Clair and the Detroit River on setlines and in pond-nets (Post, 1890; Harkness and Dymond, 1961).

It was during this same time period of the late-1800s that the Detroit River was recognized as one of the best Lake Sturgeon fisheries in the United States (Detroit Free Press 1903). Indeed, in an article titled "A Day with the Sturgeon Fishers of Fox Island," the Detroit Free Press (1903) highlighted the unique life of commercial fisherman in the late-1800s and early-1900s who lived on a small wooded island in the lower Detroit River called Fox Island and fished the surrounding wasters that were considered especially rich Sturgeon grounds. These fishermen would fish using set lines that included a special rigging of a 45.7-m (150-foot) long line stretched across the river bottom that was held in place by anchors. From this 45.7-m (150-foot) line, short 0.6-m (two-foot) lines or snubs with heavy, needle-pointed hooks would be affixed about every 0.3 m (one foot). The 47.5-m (150-foot) long lines were held up from the river bottom by ropes which ran to floats on the surface. Bottom-feeding Lake Sturgeon would come along sucking in food and get snagged on one of the heavy, needle-pointed hooks. The Detroit Free Press (1903)

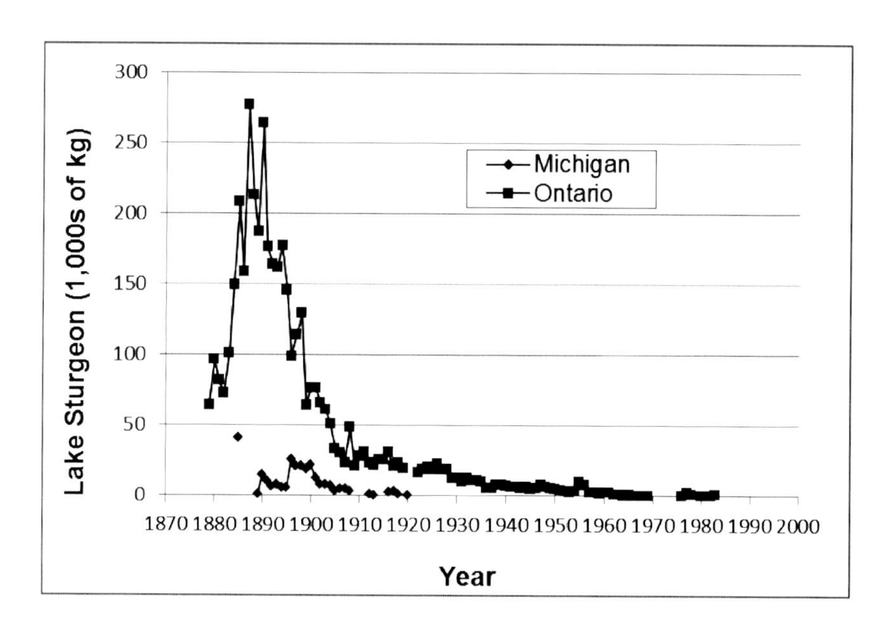

Figure 22. Lake Erie commercial fish catch in Michigan and Ontario waters, 1879-2000 (data source: U.S. Fisheries Commission Report - Fishing industry for the Great Lakes Appendix 11 to the 1926 report).

reported that passengers on boats running to Sugar and Hickory Islands could see the fishermen's floats for many kilometers stretching from the middle of Grosse Ile to the mouth of the Detroit River. Once caught, the Fox Island fishermen and others took the Lake Sturgeon to a pound or cage in the Detroit River made of boards on the sides and bottom with enough space between boards to allow water to pass through, but close enough to keep the Sturgeon from escaping. These fishermen kept the Lake Sturgeon alive in the pounds until a weekly Sturgeon buyer arrived to sell them in Detroit.

According to the Detroit Free Press (1903), Detroiters had the satisfaction of knowing that they got the best Sturgeon steaks and the best caviar in the world caught in the neighborhood of the city. During this era around the late-1800s and early-1900s, 15-20 fishermen made their headquarters on Fox Island alone, living in seven fishermen's houses or shanties (Figure 23A)(Detroit Free Press 1903). It was common for these fishermen to catch Lake Sturgeon over 1.5 m (five feet) long and weighing over 45 kg (100 pounds)(Figure 23B). The Lake Sturgeon fishery of the early-1900s was not as good as during the 1890s because of over-fishing and other factors identified above. Fresh Sturgeon sold for about 9 cents per kg (4 cents per pound) during the 1890s. By the early-1900s, with fewer Sturgeon available and continued high demand, fresh Sturgeon sold for 44 or more cents per kg (20 or more cents per pound), smoked Sturgeon sold for 55 cents per kg (25 cents per pound), and Sturgeon eggs were made into caviar and sold $1.25 per 0.45 kg (one pound) can. The Fox Island fishermen and others along the Detroit River sold their Sturgeon to buyers from Detroit during the early-1900s for about $11.00 each, no matter what the size.

It should also be noted that the Detroit River was well known for exceptional Lake Whitefish and Lake Herring (*C. artedii*) fisheries. For example, in 1872 Detroit ranked second only to Chicago, handling over 1,542,000 kg (3.4 million pounds) of fresh fish, mostly Lake Whitefish and Lake Herring (Milner, 1874). These fisheries used ponds to keep the fish alive from the fall of the year to late in the winter, when they are taken out and sold in the market at good prices. The best ponds were situated at islands in the middle of the river, like Grassy Island, where there was ample circulation of water to keep fish in healthy condition. The pond was literally an enclosure in the river, made by driving piles close together, and afterward sheathing the

A

B

Figure 23. Three Sturgeon fisherman who lived on Fox Island (A) and an over 1.5-m (five-foot) Lake Sturgeon caught off Fox Island in the Detroit River (B), circa early-1900s (photo credits: Detroit Free Press).

Figure 24. Grassy Island pond net fishery in Wyandotte, Michigan, 1874 (photo credit: U.S. Commissioner on Fish and Fisheries).

inside with planks, leaving joints of about 2 cm in width to allow the free circulation of water through the pond (Milner, 1874). At one end of the pond was a gate hinged at the bottom of the river. Figure 24 shows the Grassy Island pond net fishery on the Detroit River in 1874, complete with buildings for the fishermen, the net-house, and the storehouse. A crew of 30 men was employed in this Grassy Island fishery. Each pond would hold 25,000-40,000 Whitefish.

Lake Sturgeon populations continued to decline (Figure 22) and were nearly extirpated by the middle of the 20th Century (Roseman et al., 2011). In more recent years, the Lake Sturgeon population in Michigan has been estimated to be about one percent of its former abundance (Tody, 1974). As a result, the Lake Sturgeon is designated a "species of special concern" by the U.S. Fish and Wildlife Service, a threatened species in North America by the American Fisheries Society, a globally rare species by The Nature Conservancy, and a threatened species in the State of Michigan and Province of Ontario. Today, there is no active commercial fishery for Lake Sturgeon in the Huron-Erie corridor. Sport fishing harvest is now restricted the Michigan waters of St. Clair River and Lake St. Clair, while no Sturgeon may be possessed by anglers from Michigan or Ontario waters of the Detroit River (Great Lakes Fishery Commission, 2003; Michigan Department of Natural Resources, 2005; Ontario Ministry of Natural Resources, 2005).

Probably the most dramatic example of Lake Sturgeon spawning habitat destruction was the construction of the shipping channels in the Detroit River. Millions of tonnes of limestone bedrock, cobble, and gravel were removed from the lower Detroit River to create the shipping channels to support commerce. These rock and gravel substrates provided critical spawning habitats for lithophilic (i.e. rock loving) broadcast spawning fishes like Lake Sturgeon, Lake Whitefish, Walleye (*Sander vitreus*), and others. This navigational work first began in 1874 with the construction of the Limekiln Crossing and continued on through the early-1900s with the construction of the Livingstone and Amherstburg channels (Bennion and Manny, 2011). The construction of the Livingstone Channel in the early-1900s was particularly damaging to spawning habitat (Figure 25). A 19-km channel (minimum width 91 m; minimum depth: 6.7 m) was created in a limestone bedrock sill at the mouth of the Detroit River to facilitate shipping, greatly altering the river's hydrology and destroying many critical fish spawning grounds (U.S. Bureau of Fisheries, 1917; Manny et al., 1988).

Restoration Efforts

From the 1970s to 1999 no Lake Sturgeon spawning was reported in the Detroit River which, as noted above, was one of the most productive Sturgeon spawning grounds in the United States. Again, it is notable that both Michigan and Ontario have closed the entire Detroit River to possession of Lake Sturgeon, to conserve this potentially-important sport and commercial fish species to help further recovery of the Detroit River population.

In 2001, however, Lake Sturgeon spawning was documented on a coal cinder pile near Zug Island in the Detroit River for the first time in over 20 years (Caswell et al., 2004). In response to the discovery of Lake Sturgeon spawning, scientists performed monitoring and research to determine the extent of the Sturgeon population in the Detroit River, including possible spawning locations and success rates. From 2000 to 2002, they fished with set-lines for 741 days total, while the river was ice free, and only caught 85 Lake Sturgeon.

Figure 25. Construction of Livingstone Channel in the Detroit River during the early-1900s (photo credit: Detroit Publishing Company Collection, Library of Congress).

If this same experiment had been conducted in the late-1800s, probably more than 1,000 Lake Sturgeon would have been captured. Fishery biologists and managers concluded that in the 2000s Lake Sturgeon reproduction was now more limited by habitat. The theory put forward was that if Lake Sturgeon spawning habitat could be restored in the right areas, with the proper conditions, and in close proximity to critical nursery areas, the population would grow and hopefully recover.

As a result, Lake Sturgeon spawning habitat was created in the Detroit River off:

- McKee Park in Windsor, Ontario in 2003,
- Belle Isle in Detroit, Michigan in 2004;
- Fort Malden in Amherstburg, Ontario in 2004; and

- BASF property along the Trenton Channel in Riverview, Michigan in 2008.

Disappointingly, no Lake Sturgeon reproduction has been documented at any of these four sites. However, at the Belle Isle reef, pre-construction assessment efforts of the area resulted in the capture of three species of fish with no spawning documented. Post-construction assessment of the Belle Isle reef in 2005 and 2006 identified 16 species of native fish spawning at the site (including Lake Whitefish, a first in over 90 years) with three other species using the site, including two state listed species, Lake Sturgeon and the Northern Madtom (*Noturus stigmosus*) (Manny, 2006).

Additional fishery surveys and assessments performed in the Detroit River during 2006–2008 showed that adult Lake Sturgeon were present and that spawning conditions were favorable (Roseman et al., 2011). Aided with this knowledge, fishery managers again mobilized to build a Sturgeon spawning reef off Fighting Island where:

- they were known to historically spawn;
- spawning conditions were met;
- adult Lake Sturgeon were frequently captured; and
- nursery habitat was in close proximity.

Based on available scientific evidence, fishery biologists and resource managers from 17 different Canadian and U.S. organizations agreed to collaborate on designing and constructing a Sturgeon spawning reef off the northeastern portion of Fighting Island. This binational group came together and reached agreement on a concept proposal for the Fighting Island reef. The concept proposal was then modified by different organizations and agencies in both Canada and the U.S., and expanded into full proposals that were submitted to U.S. and Canadian nonprofit organizations, foundations, governmental agencies, and corporations. In total, over $250,000 was raised in Canada and the U.S. for construction of the Fighting Island reef. The money was pooled and transferred to Essex Region Conservation Authority for reef construction in 2008, representing the first ever fish habitat restoration project in the Great Lakes funded with both U.S. and Canadian funds.

A 3,300 m^2 Sturgeon spawning reef was constructed at the upstream end of Fighting Island in the fall of 2008. The reef design included four different types of substrate (i.e. limestone "shot rock," limestone "sorted rock," rounded rock, and a mixture of the three) with three replicates each (Roseman et al., 2011). In total, 12 spawning beds were constructed across the entire width of the Fighting Island channel.

Lake Sturgeon showed an immediate response by spawning during the first spring season following reef construction (Figure 26). During May and June 2009, ripe, spawning-ready adults, viable eggs, and larvae were collected off the reef, documenting natural reproduction in the Canadian portion of the Detroit River for the first time in 30 years (Roseman et al., 2011). In 2010, no sampling was conducted in the river for Sturgeon larvae, but eggs and spawning-ready adults were collected there, and three age-0 juvenile Lake Sturgeon were captured in bottom trawls fished downstream of the reef during July (Roseman et al., 2011). Interestingly, spawning Lake Sturgeon showed no preference among the four different reef substrate types, but did demonstrate a preference for the Fighting Island side of the channel, where faster water velocities occurred (Roseman et al., 2011).

It should be noted that Walleye and Lake Whitefish also spawned on the Fighting Island reef, suggesting that the constructed spawning habitat may also be enhancing populations of these high-value fish. Other exciting news at the Fighting Island reef was the collection of several Northern Madtom (*Noturus stigmosus*), a fish that is endangered in Michigan and Ontario and never before confirmed in the mid-reaches of the Detroit River. This is important because it demonstrates that this type of habitat restoration is of value for sustaining threatened and endangered native fish populations.

Based on the evidence of natural reproduction of Lake Sturgeon on the recently constructed artificial Fighting Island reef, fishery biologists concluded that further expansion of the reef could enhance reproduction and early life history survival of Lake Sturgeon in the Detroit River (Roseman et al., 2011). Efforts are now underway to expand the size of the Fighting Island reef and to construct several more reefs on the U.S. side of the Detroit River.

Fishery biologists from U.S. Fish and Wildlife Service also performed mark-and-recapture studies during 2003-2012 in the Detroit River. These studies involved capturing 214 Lake Sturgeon in the river, tagging them and returning them to the river, recapturing some of them in subsequent fishing, and using the ratio of marked to unmarked fish in the catch as a basis for estimating the size of the population. A total 214 Lake Sturgeon was tagged in the Detroit River (Chiotti et al., 2012). These mark-and-recapture data were input into models which estimated the size of the Lake Sturgeon population in the Detroit River at 4,068 individuals with a 95% confidence interval of 869-7,269 (Chiotti et al., 2012). Additional assessments are underway to track and understand the population status. U.S. Fish and Wildlife Service fishery biologist Jim Boase noted:

> *Based on our fishery assessment work performed in 2003-2012, our estimate of over 4,000 Lake Sturgeon living in the Detroit River is wonderful news. This is a much larger population than we thought and a very positive sign for our U.S.-Canada efforts to restore the Lake Sturgeon population. Further, with confirmation of natural reproduction of Lake Sturgeon at both the Zug Island and Fighting Island sites, we feel that the future of this threatened species is very promising.*

Figure 26. Lake Sturgeon caught off the Fighting Island reef in 2011 (photo credit: U.S. Fish and Wildlife Service).

Concluding Thoughts

The story of Lake Sturgeon has been written in a number of chapters, including revered by Native Americans, a nuisance to commercial fisherman, sought after as the "King of Caviar," and recovering species championed as an indicator of ecosystem health. Clearly, this recovery story and trend are very encouraging.

The successful reproduction of Lake Sturgeon in the Detroit River after a 30-year absence is, in part, a result of 40 years of water pollution prevention and control programs in the Detroit/Windsor metropolitan areas (Roseman et al., 2011). However, much remains to be done to ensure that this threatened species achieves a sustainable population level. Recommended actions include:

- enforcing conservative Sturgeon harvest regulations throughout the Huron-Erie corridor;
- enhancing and protecting Sturgeon spawning and nursery habitat in the Detroit River (i.e. expanding the Fighting Island reef and constructing additional ones on the U.S. side);
- undertaking regular Sturgeon population assessments, including the use of telemetry, to decipher spawning sites and movement patterns;
- monitoring Lake Sturgeon population dynamics and trends, and factors affecting them;
- controlling pollution from combined sewer overflows and other sources within the watersheds; and
- expanding the St. Clair-Detroit River Sturgeon for Tomorrow nonprofit organization to build the capacity of governments to preserve and protect a sustainable Lake Sturgeon population and fishery.

Further, the Fighting Island Sturgeon spawning reef is an excellent example of strengthening the science-management linkage and achieving compelling ecosystem results. Continued priority must be placed on achieving a close coupling of science and management to ensure both cost- and ecosystem-effective decisions. Greater public understanding and support is needed to:

- protect this species as an important member of the Great Lakes' food web;
- recognize its historical, cultural, and future food value to humans; and
- remember that if the Detroit River environment is cleaner for Lake Sturgeon, it is cleaner for humans.

Literature Cited

Bein, L., 2012. In the Archives: From Cordwood to Cavier. The Ann Arbor Chronicle, February 28th Edition, Ann Arbor, Michigan, USA. http://annarborchronicle.com/2012/02/28/in-the-archives-from-cordwood-to-caviar/ (April 2013).

Bennion, D.H., Manny, M.A., 2011. Construction of shipping channels in the Detroit River: History and environmental consequences. Scientific Investigations Report 2011-5122, U.S. Geological Survey, Reston, Virginia, USA.

Bogue, M.B., 2000. *Fishing the Great Lakes: An Environmental History, 1783-1933.* University of Wisconsin Press, Madison, Wisconsin, USA.

Caswell, N.M., Peterson, D.L., Manny, B.A., Kennedy, G.W., 2004. Spawning by Lake Sturgeon (*Acipenser fulvescens*) in the Detroit River. Journal of Applied Ichthyology 20, 1-6.

Chiotti, J., Mohr, L., Thomas, M., Boase, J., Manny, B., 2012. Lake Sturgeon population demographics in the St. Clair-Detroit River System, 1996-2012. American Fisheries Society 142nd Annual Meeting (August 19), Minneapolis-St. Paul, Minnesota, USA.

Cornell, G.L., 2003. American Indians at Wawiiatanong: An early american history of indigenous peoples at Detroit. In, J.H. Hartig (Ed.), *Honoring our Detroit River: Caring for our Home*, pp. 9-22. Cranbrook Institute of Science, Bloomfield Hills, Michigan, USA.

Detroit Free Press, 1903. *A day with the Sturgeon fishers of Fox Island.* Detroit, Michigan, USA.

Duglas, H., Keeney, S., Cahoon, J.A., 1877. Commercial fishing logs from Pointe Mouillee, 1876-1877. Marshlands Museum, Lake Erie Metropark, Huron Clinton Metropolitan Authority, Brownstown, Michigan, USA.

Givens-McGowan, K., 2003.The Wyandot and the River. In: J.H. Hartig (Ed.), *Honoring our Detroit River: Caring for our Home*, pp. 23-34. Cranbrook Institute of Science, Bloomfield Hills, Michigan, USA.

Great Lakes Fishery Commission, 2003. Lake Sturgeon in the Great Lakes; Lake Sturgeon, the giant of the Great Lakes. Great Lake Fisheries Commission, Ann Arbor, Michigan, USA.

Harkness, W. J. K., Dymond, J.R., 1961. The Lake Sturgeon. Ontario Department of Lands and Forests, Fish and Wildlife Branch, Peterborough, Ontario, Canada.

Hay-Chmielewski, E., Whelan, G., 1997. Lake Sturgeon rehabilitation strategy. Fish Division, Michigan Department of Natural Resources, Special Report Number 18, Ann Arbor, Michigan, USA.

Lanman, J.H., 1839. *History of Michigan.* E. French, New York, New York, USA.

Longfellow, H.W., 1855. The Song of Hiawatha. eBooks@Adelaide, The University of Adelaide Library, University of Adelaide, South Australia. http://ebooks.adelaide.edu.au/l/longfellow/henry_wadsworth/song_of_hiawatha/index.html (March 2012).

Manny, B.A., 2006. Monitoring element of the Belle Isle/Detroit River Sturgeon Habitat Restoration, Monitoring, and Education Project. Report to Michigan Sea Grant, Ann Arbor, Michigan, USA.

Manny, B.A., Edsall, T.A., Jaworski, E. 1988. The Detroit River, Michigan – An ecological profile. U.S. Fish and Wildlife Service Biological Report 85(7.17), Ann Arbor, Michigan, USA.

Michigan Department of Natural Resources, 2005. Fishing Regulations, Lake Sturgeon (PDF).

Milner, J. W., 1874. Report on the fisheries of the Great Lakes: The result of inquiries prosecuted in 1871 and 1872. Report of the U.S. Commissioner of Fish and Fisheries for 1872 and 1873. 2:1-78, Washington, D.C., USA.

OMNR (Ontario Ministry of Natural Resources), 2005. Fishing regulations for the province of Ontario.

Post, H., 1890. The Sturgeon; some experiments in hatching. Trans. Amer. Fish. Soc. 19, 36-40.

Roseman, E.F., Manny, B., Boase, J., Child, M., Kennedy, G., Craig, J., Soper, K., Drouin, R., 2011. Lake Sturgeon response to a spawning reef constructed in the Detroit River. J. Appl. Ichthyol. 27 (Suppl. 2), 66–76.

Scott, W.B., Crossman, E.J., 1973. *Freshwater fishes of Canada*. Fisheries Research Board of Canada. Ottawa, Ontario, Canada.

Tody, W.H., 1974.Whitefish, Sturgeon, and the early Michigan commercial fishery. In: Michigan Fisheries Centennial Report 1873-1973, pp. 45-60, Michigan Department of Natural Resources, Lansing, Michigan, USA.

U.S. Bureau of Fisheries, 1917. Report of the U.S. commissioner of fisheries for fiscal year ended June 30, 1916. Bureau of Fisheries Doc. 836, Washington, D.C., USA.

U.S. Fish and Wildlife Service, 2012. Great Lakes Lake Sturgeon Web Site. Alpena, Michigan, USA. http://www.fws.gov/midwest/Sturgeon/biology.htm (May 2012).

CHAPTER 8

Citizen Science and Stewardship

Whether it's counting birds, listening for frogs, spotting salamanders, collecting butterflies or dragonflies, identifying aquatic invertebrates collected from sediments, gauging stream flow, or measuring water quality — there is an increasing and ardent need for citizens to participate in science. Further, natural resource and environmental management agencies are frequently challenged with limited resources to properly collect and analyze data to adequately inform natural resource decision-making. Therefore, need for and interest in citizen science is growing.

Citizen science is scientific research and monitoring conducted, in whole or in part, by amateur or nonprofessional scientists. One easy way to think of it is as public participation in scientific research and monitoring. Formally, citizen science has been defined as the systematic collection and analysis of data, development of technology, testing of natural phenomena, and the dissemination of these activities by researchers on a primarily avocational basis. For example, a fundamental premise of citizen science is that anyone who watches birds, from backyards to city streets to remote forests, can help make a contribution of our scientific understanding and management of birds, their habits, and their habitats.

But it is not just scientists that benefit. Being an amateur scientist can be very rewarding as volunteers gain hands-on experience with scientific methods, learn about ecological principles and practices, make a contribution to the growth and expansion of scientific knowledge, often help solve environmental and natural resource problems, and network with others who have a common interest (Yung, 2007). Many volunteers welcome the opportunity to just be outside and participate in fieldwork.

Similar to the growing interest in citizen science, there is growing interest in stewardship. In the conservation field, stewardship is defined as the careful and responsible management of natural resources entrusted to one's care. Citizen and community-based stewardship is public participation in fish and wildlife conservation. It is based on the premise that fish and wildlife conservation is a responsibility shared among citizens, governments, and other stakeholders. Indeed, this tenet of shared responsibility is even given in the mission statement of the U.S. Fish and Wildlife Service (USFWS):

> *The mission of the U.S. Fish and Wildlife Service is working with others to conserve, protect, and enhance fish, wildlife, plants, and their habitats for the continuing benefit of the American people.*

A commitment to citizen science and stewardship is also given by the USFWS in the Comprehensive Conservation Plan of the Detroit River International Wildlife Refuge (DRIWR) that calls for establishing partnerships and using volunteers to identify and monitor wildlife, and to undertake stewardship (U.S. Fish and Wildlife Service, 2005). Experience has shown that participation in such citizen science and stewardship can promote scientific and conservation literacy, and can help develop a stewardship ethic in the citizenry. This chapter will share examples of how governments are working with citizens and other partners to build the capacity to monitor and survey fish and wildlife populations, and to be good stewards of the refuge's amazing natural resources.

Detroit River Hawk Watch and Holiday Beach Migration Observatory

It was a beautiful fall morning when we arrived at Lake Erie Metropark at the mouth of the Detroit River to go for a walk. It was clear and the warmth of the sun felt like a warm blanket on a cold winter morning. As I looked up to appreciate the beauty of this glorious fall morning, I noticed thousands of birds soaring and swirling in the air. As I continued my walk, watching this display of avian flying mastery, I came upon a naturalist who was eager to explain this phenomenon and share her passion for these migrating birds.

Each fall it is amazing to watch hundreds of thousands of raptors migrate across the lower Detroit River from their eastern Canadian breeding grounds and head south to their wintering grounds, some as far away as South America. These raptors are birds of prey that hunt food primarily by flight, using their keen senses, especially vision. The term raptor comes from the Latin word *rapere* that means to seize or take by force. Examples of raptors include Bald Eagles, Peregrine Falcons, Osprey, hawks, Turkey Vultures, and others.

Raptors use thermals to varying degrees to aid in their migration to save energy for their long journey. Thermals are rising columns of warm air that are caused by the heating of the earth by the sun. A raptor will literally ride up on a thermal, set its wings, and glide downward to the next thermal using little energy. Thermals do not form over water, so as these raptors head south and come upon lakes Erie and Ontario, they have one of two choices: fly east around Lake Ontario or fly west around Lake Erie. Those that move west follow the northern shore of Lake Erie until they reach the mouth of the Detroit River. Turning back is not an option, so they are forced to cross the 6-km (4-mile) river mouth near Holiday Beach in southwest Ontario to southeast Michigan in the vicinity of Lake Erie Metropark, Humbug Marsh Unit of the DRIWR, and Pointe Mouillee State Game Area. Lake Erie Metropark is the primary location for the Detroit River Hawk Watch program and hosts an annual Hawk Fest that attracts more than 5,000 birders. These raptors lose altitude as they cross the lower Detroit River, making it easier for them to be observed. To give you a feel of the magnitude of hawk migrations through the lower Detroit River, dedicated volunteers recorded an astounding 190,121 Broad-Winged Hawks on September 17, 2011 at Lake Erie Metropark, representing the third highest single-day total recorded at this site (Stein, 2011).

This predictable hawk migration phenomenon provides USFWS with a unique opportunity to involve birders in the systematic collection of annual raptor migration data. The USFWS, its friends' organization called the International Wildlife Refuge Alliance (IWRA), and avid birders undertake Detroit River Hawk Watch on an annual basis to: systematically count hawks during their migration season; review and analyze the data; prepare summary reports; and disseminate the data and findings to both managers and the public. A paid counter is employed to work with volunteer counters to collect the data (Figure 27). USFWS oversees the Detroit River Hawk Watch monitoring effort, including quality assurance/quality control and report preparation. IWRA helps with fund-raising, hires the annual paid counter, recruits volunteers, maintains a Detroit River Hawk Watch website, and helps coordinate outreach at the annual Hawkfest at Lake Erie Metropark (September of each year). A Detroit River Hawk Watch Advisory Committee has also been established to provide citizen advice to both USFWS and IWRA on all aspects of Detroit River Hawk Watch. Such hawk watch data are invaluable in tracking raptor migrations and providing an early warning signal for changes in trends (Table 12). Further, these data are also entered into the Hawk Migration Association of North America database to help understand raptor population status and trends on a continental scale.

Figure 27. Volunteers participate in Detroit River Hawk Watch at Lake Erie Metropark (photo credit: U.S. Fish and Wildlife Service).

Similar to Detroit River Hawk Watch in the U.S., a nonprofit organization named Holiday Beach Migration Observatory runs a long-term monitoring program for migratory bird populations during the spring and fall seasons at the Holiday Beach Conservation Area in southwest Ontario. Holiday Beach Conservation Areas is owned by Essex Region Conservation Authority, is listed on the Canadian registry of lands for the DRIWR, and located right across the Detroit River from Lake Erie Metropark. Like Detroit River Hawk Watch, the Holiday Beach Migration Observatory uses volunteers and a paid counter to document the migration of raptors and other species, and all data are entered in the Hawk Migration Association of North America database. This citizen science program has a 40-year history and includes considerable education and outreach activities, including an annual Festival of Hawks in September. Clearly, these two sister monitoring programs are an outstanding example of using citizen science to provide mission-critical data to managers in a scientifically-defensible manner.

Table 12. Selected examples of 2012 raptor counts from Detroit River Hawk Watch and percent deviation from the previous 14-year average (Stein et al., 2013).

Species	Period of Record	Total 2012 Count (Percent Deviation from Previous 14-year Average)
Turkey Vulture (*Cathartes aura)*	1998-2012	43,285 (-5.2%)
Bald Eagle (*Haliaeetus leucocephalis)*	1998-2012	222 (+53%)
Northern Harrier (*Circus cyaneus*)	1998-2012	248 (-50.1%)
Sharp-Shinned Hawk (*Accipiter striatus*)	1998-2012	3,590 (-48.5%)
Northern Goshawk (*Accipiter gentilis*)	1998-2012	18 (-27.2%)
Cooper's Hawk (*Accipiter cooperii*)	1998-2012	468 (-13%)
Red-Tailed Hawk (*Buteo jamaicensis*)	1998-2012	2,986 (-40.3%)
Broad-Winged Hawk (*Buteo platypterus*)	1998-2012	40,923 (-36.2%)

Marsh Bird Monitoring

Picture a group of birds that are colored for camouflage, have secretive behaviors, and reside in places that are not very accessible to humans. This pretty much describes marsh-dwelling water birds or what birders call marsh birds. Because of the above characteristics, not much is known about their population dynamics or trends. Yet, marsh birds face a host of conservation concerns in the Great Lakes Basin, including habitat loss and fragmentation, decreasing water levels, urban encroachment, wetland contamination by nonpoint source pollution, consumptive uses like hunting for snipe, rails, and gallinules, invasion of non-native plant species, and even marsh management practices like herbicide treatment of invasive *Phragmites*. Examples of marsh birds include: Marsh Wrens (*Cistothorus palustris*), rails, bitterns, grebes, gallinules, and snipe.

What we do know is that marsh birds are important components of wetlands and some are of conservation concern. For example, the Yellow Rail (*Corturnicops noveboracensis)* and the Least Bittern (*Ixobrychus exilis)* are designated as threatened species in Michigan, and the American Bittern (*Botaurus lentiginosus*) and Marsh Wren are considered "special concern" species in Michigan. Marsh birds are seldom heard because they vocalize infrequently and prefer to inhabit inaccessible wetland habitats. Despite their inaccessibility, marsh birds can be an indicator of wetland health.

In the DRIWR, students, volunteers, and the Michigan Natural Features Inventory are working with the Refuge Biologist to survey marsh bird use of the newly restored wetlands within the Brancheau and Gibraltar Wetlands units. Marsh bird data, including photo points at each station, are systematically collected following the Standardized North American Marsh Bird Monitoring Protocols (Conway, 2009) and entered into the refuge's Geographical Information System. These protocols provide standardized methods for documenting presence, distribution, and density of focal species. In essence, citizen scientists play recordings of marsh birds and then listen for a response, increasing the chances of detecting a focal species. Also included in the marsh bird surveys are descriptions of habitat characteristics at survey sites, such as percent open water and dominant vegetation. This information is useful in understanding species-habitat relationships.

These marsh bird data are used by staff of the DRIWR to help document the effectiveness of the recent restoration of 27.1 ha (67 acres) of wetlands on former agricultural land in the Brancheau Unit and the mitigation of 106 acres of wetlands at the Gibraltar Wetlands Unit. These data will also be used to help document the effects of additional marsh management actions in the future. Marsh birds surveyed in the Brancheau and Gibraltar Wetlands units include: Sora, Virginia Rail, Common Moorhen, Marsh Wren, Least Bittern, and American Coot.

The partnership with the Michigan Natural Features Inventory is particularly noteworthy because of the long-term commitment to marsh bird monitoring. The Michigan Natural Features Inventory goal of is to evaluate long-term trends in marsh bird abundance and distribution at selected sites throughout Michigan (including the Gibraltar Wetlands Unit), facilitate conservation planning, and address knowledge gaps about marsh bird habitat associations. Each survey area or “route” consists of about 4-8 points at which surveys are conducted three times per season between early May and mid-June. These surveys are conducted at the refuge’s Gibraltar Wetlands Unit by a volunteer and paid Michigan Natural Features Inventory staff on an annual basis.

Christmas Bird Count

In 1900, American ornithologist and Audubon Society Officer Frank Chapman asked birders across North America to head out on Christmas Day to count the birds in their home towns and submit the results as a "Christmas Bird Census." This first “Christmas Bird Census” has now become an annual holiday tradition called the Christmas Bird Count, representing the longest running citizen science survey in the world and the longest running database in ornithology. Tens of thousands of participants are involved in over 2,000 locations across Canada, the United States, Latin America, and the Caribbean (Yung, 2007). These bird observations have been amassed into a huge database that reflects the distribution and numbers of winter birds over time.

The Detroit River Christmas Bird Count was established in 1978 as a U.S.-Canada endeavor to count overwintering birds (Craves, 2007). The area covered is a circle centered on Detroit, encompassing parts of both Wayne County in Michigan and Essex County in Ontario.

The count circle consists of mostly industrial, urban, and residential areas, but includes some natural areas, such as the Ojibway Prairie in Windsor and wetlands along the Detroit River that are critical for bird populations. On the Canadian side a number of Christmas Bird Counts are coordinated each year by Bird Studies Canada in cooperation with numerous volunteers. Christmas Bird Counts are conducted along the upper reaches of the Detroit River, at the Holiday Beach Conservation Area, and at Point Pelee National Park.

These Christmas Bird Count data represent a one-day snapshot of the status and distribution of early winter bird populations recorded between December 14th and January 5th continent-wide. Each individual count has a relatively long history and is vital to conservation efforts. These population data can act as "early warning signals" of habitat change in a count area by examining trends of certain species or suites of species that depend on specific habitats. Also, local trends in bird populations can signal an immediate environmental threat, such as toxic substances' contamination, and can help document the appearance, presence, or increase of non-native or exotic species, which often have profound impacts on ecosystem health.

Citizen Involvement in Common Tern (*Sterna hirundo*) Restoration and Monitoring

Can you imagine a group of students and other volunteers getting excited about being attacked by a colonial water bird while trying to restore its population? Well, that is precisely what happened under the leadership of a high school teacher from Southgate Anderson High School named Bruce Szczchowski and the Refuge Biologist named Greg Norwood, who both went on to get their Master's Degree on this important work. These two passionate biologists first worked with citizen stewards to restore or enhance Common Tern habitat at two Grosse Ile bridges and DTE Energy's Rouge Power Plant (Figure 28). Although these efforts led to no significant expansion of the Common Tern population, they did result in considerable practical knowledge that was applied in re-creating Common Tern nesting habitat on one of the most important historical nesting locations along the river corridor – Belle Isle.

Belle Isle is a 397-ha (980-acre) island park in the Detroit River. In the early-1960s, as many as 1,200 pairs of Common Terns nested on the eastern tip of Belle Isle. During this time, park management deemed the nesting habitat to be a nuisance and destroyed it to create turf-grass lawn. Based on the citizen science data and information collected by Bruce Szczchowski, Greg Norwood, and their students and volunteers, the overall Common Tern restoration effort was expanded with Detroit Zoo to experiment with the recreation of Common Tern nesting habitat on the eastern tip of Belle Isle. In 2008 and 2010, 799 m^2 (8,600 square feet) of sand and crushed limestone were placed on the eastern tip in the very same area where the nearly 1,200 nesting pairs of common terns nested in the early-1960s. Over a two-year period, project partners and students from Detroiters for Environmental Justice and Southgate Anderson High School helped recreate nesting habitat using crushed limestone, removed invasive vegetation, planted native plants beneficial to incubating terns and their chicks, erected a sheep fence to exclude predators during the nesting season, and placed 100 decoys and a solar-powered sound

Figure 28. Citizen scientists help restore Common Tern nesting habitat at the Grosse Ile free bridge (photo credit: U.S. Fish and Wildlife Service).

system to broadcast non-aggressive calls to help attract this threatened migratory bird species back to Belle Isle. In 2011, citizen scientists observed that the colony was actively defended by Common Terns from mid-April to mid-July. Greg Norwood noted that:

> *One muggy July day in 2010, a Common Tern pecked at our heads while we were monitoring the site. This feisty behavior was precisely what we were hoping to see because that means that the Common Terns had established an affinity for the site and were fiercely defending it as their breeding habitat.*

In 2012, two Common Terns were confirmed to have fledged from the restored habitat on Belle Isle, representing the first such fledging success since the 1960s. This project was an excellent example of both citizen science and stewardship, and provided a unique opportunity in citizens and students to be involved in a compelling conservation experience to restore a threatened species.

Tree Planting and Stewardship in Essex County, Ontario

Within the Canadian Priority Natural Area (Chapter 2), Essex Region Conservation Authority (ERCA) owns or manages 19 publicly accessible properties totaling more than 1,000 hectares (4,000 acres) of woodlands, marshes, and shoreline areas in the region. The United Nations has recommended that, at a minimum, 12% of any region must remain in its natural state in order to be healthy and sustainable (United Nations World Commission on Environment and Development, 1987). In Essex County, only 8.5% of the current land base is considered natural areas. To address this deficiency, ERCA has partnered with public and private landowners to implement large-scale naturalization projects ranging from 0.4 ha (one acre) to more than 40 ha (100 acres).

Natural area restoration projects are implemented on lands owned by individuals, businesses, or municipalities. The majority are undertaken in partnership with rural private landowners who wish to reforest or naturalize parcels of land, or complete projects that enhance water quality. The types of restoration projects included:

- conventional tree planting;
- pit and mound forest restoration;

- wetland enhancement/construction;
- tallgrass prairie/wildflower meadow restoration;
- vegetated buffer strips; and
- soil erosion control structures (e.g. rock chutes, header tile retrofits, septic system upgrades).

ECRA has literally worked with thousands of landowners on these projects that have improved aesthetics, enhanced wildlife habitats, and increased property values. The broader health benefits are cleaner water, healthier air, and increased natural areas coverage. Of particular note is their forestry program that has planted almost six million trees in the Essex County, most with volunteers. Trees provide natural habitat for wildlife, protect soil from eroding, and improve water quality. Trees have been demonstrated to add up to 15% to the value of a home, and if planted in strategic locations, can reduce home heating and cooling requirements. ERCA also has a very active seed collection program. Each fall, ERCA collects seeds from local trees which are then propagated into seedlings which get planted in future years. Science has shown that trees, grown from local stock, are most resilient to the climate and soil conditions of that particular area, thereby increasing their survivability.

Stewards of Humbug Marsh

Although Humbug Marsh is the last kilometer of natural shoreline on the U.S. mainland of the Detroit River and Michigan's only "Wetland of International Importance" under the Ramsar Convention, it has a storied past, including:

- being a private hunt club that at one time even built dikes to enclose the marsh to maximize waterfowl production,
- serving as a storage yard for military vehicles during World War II,
- being farmed,
- providing pasture land for sheep grazing, and
- portions even being brush-hogged for possible a future housing development that never materialized.

Despite these perturbations, Humbug Marsh has been resilient and still exhibits exceptional ecological characteristics, including its mesic flatwoods habitat. In laypersons' terms that means that Humbug Marsh is an oak-hickory forest that is classified as "mesophytic", meaning that plants species are adapted to neither particularly dry nor particularly wet conditions. In addition to some of the perturbations identified above, Humbug Marsh also shows impacts of high deer browse, invasive species, and fire suppression.

Therefore, there is a unique opportunity for citizens to participate in stewardship activities for habitat restoration and enhancement. Stewardship volunteers, under the leadership of biologist Greg Norwood, are slowly and incrementally removing invasive plants like buckthorn, honeysuckle, and Garlic Mustard, exposing the understory to more light, and enhancing soil development and soil moisture retention.

Over the past decades, invasive plants have come to dominate portions of Humbug Marsh. These invasive plants promote more invasive plants by limiting the amount of sunlight that reaches the ground and contributing to poorly aggregated soils and poor moisture conditions. It is precisely these conditions that give invasive species like buckthorn, honeysuckle, and Garlic Mustard competitive advantage over others.

This group of hearty stewards is now working hard to remove invasive plants, thereby opening the ground to more light and promoting better soil development and greater soil moisture content. This, in turn, is contributing to the germination of native grasses like blue joint and native sedges from a seed bank that still exists in Humbug Marsh. It is these native grasses and sedges that are improving the soil conditions and improving the soil moisture content through natural ecological processes. Native grass and sedge root systems grow into the upper soil layer, die continuously, and partially decompose, enhancing soil aggregation and soil moisture content. Soil aggregation allows more infiltration of precipitation.

For the past several years, a highly committed group of men and women have been cutting buckthorn, removing honeysuckle, and pulling Garlic Mustard as part of larger efforts to restore more native coastal habitat. This group works every Wednesday from 9:00 AM till noon, including during winter (Figure 29). Most of the volunteers are retirees from a variety of backgrounds like teaching, law enforcement, utility companies, and the automotive industry. But they all share a love of the out-of-doors and are committed to becoming stewards of Michigan's only Wetland of International Importance designated under the International Ramsar Convention – Humbug Marsh. Group leader Dick Skoglund of Trenton, had this to say about his volunteer experience at Humbug Marsh:

> *Volunteering for stewardship activities is an incredible opportunity to make a tangible contribution to North America's only international wildlife refuge. It gives me enormous personal satisfaction.*

Figure 29. Committed volunteers from the Refuge's Stewardship Commiittee remove invasive buckthorn at Humbug Marsh, winter 2013 (photo credit: International Wildlife Refuge Alliance).

Their results from 2012 alone were amazing, including the removal of buckthorn, honeysuckle, and Garlic Mustard from 10 acres within the Humbug Marsh Unit, the killing of 130 invasive European black alders from along the Detroit River shoreline, and the broadcasting of native seed collected from high quality habitats within Humbug Marsh. Refuge staff and the IWRA are now working to expand these stewardship efforts by adding additional volunteers to this Wednesday morning group and/or by establishing an additional stewardship group working at different times.

Shoreline Restoration at the Refuge's Gibraltar Bay Unit

Most projects need a catalyst and champion, and the shoreline restoration at the Gibraltar Bay Unit was no exception. Dr. Bruce Jones, a retired dentist from Grosse Ile Township and board member of Grosse Ile Nature and Land Conservancy, was passionate about environmental education in schools and about developing a stewardship ethic in the next generation of community leaders. In 2002, he convinced the Grosse Ile Nature and Land Conservancy and a number of partners of the need to: restore a more natural shoreline at the Gibraltar Bay Unit, promote environmental education through the project to local schools, and encourage a stewardship ethic. Dr. Jones recruited Nativescape, LLC to help develop a restoration plan and the Greater Detroit American Heritage River Initiative to help secure necessary funding.

One man's vision became a community reality. In the summer of 2003 the first phase of shoreline restoration was undertaken. The first task was to remove decades of debris that was dumped along the shoreline and that accumulated from wind and wave action. The solution was to recruit the Navy Seabees to undertake the debris removal as a training exercise. The next step included rehabilitating 85 meters (280 feet) of bay shoreline using soft engineering techniques. This shoreline was originally part of an old Nike Missile base. The missile base was created in the 1950s by filling in a shallow section of Gibraltar Bay on the Detroit River. The base was closed in the early-1960s and was dismantled in the early-1990s. Therefore, the shoreline has not been natural for some time.

The design of a more natural shoreline included use of a new technique, where biodegradable plastic tubes (i.e. fibersock or "soil sock") were used to stabilize and restore the shoreline. Clean-composted recycled yard waste and small stone was pneumatically pumped into the tubes along with a mixture of native emergent plant seeds. The fibersock was then placed along the shoreline edge and anchored in place. The compost mixture was then back filled into the space between the tube and the old shoreline creating a new aquatic shelf. A geofabric blanket was then placed over this back fill to stabilize the area until the plants grew. A group of Grosse Ile high school students planted about 1,400 emergent plant plugs to help create the aquatic shelf.

Over time, deposition of material occurred in front of the fibersock. Eventually, the fibersock degraded and resulted in a new natural shoreline with native emergent plants. The second phase of this project was undertaken in 2004-2005 and involved completing the restoration of the remaining portion of the shoreline using this soft engineering technique. Total cost of both phases was $80,000. Other partners in the project included: Metropolitan Affairs Coalition, U.S. Fish and Wildlife Service, Great Lakes Basin Program for Soil Erosion and Sediment Control, Grosse Ile Schools, Downriver Community Conference, Friends of the Detroit River, Michigan Sea Grant, and the Detroit/Wayne County Port Authority.

Prior to shoreline restoration, the plant community had limited diversity and was dominated by invasive non-native species, including Giant Reed (*Phragmites australis*), Purple Loosestrife (*Lythrum salicaria*) and reed Canary Grass (*Phalaris arundinacea*). There were a limited number of native plant species. A macro-invertebrate study also showed a limited number of species using the site and Bull Frogs (*Rana catesbeianaa*) were absent.

After shoreline restoration, a more diverse plant community was present, dominated by native species that are now seeding surrounding areas (i.e. Cupplant – *Silphium perfoliatum*), Swamp Rose Mallow (*Hibiscus palustris*), and Cardinal Flower (*Lobelia cardinalis*). None of these species were present before restoration. A macro-invertebrate study performed by students, teachers, and biologists showed an increase in diversity. Bull Frogs have returned to site. Students and other volunteers were instrumental in both planting aquatic plugs and monitoring conditions. All feedback from students and volunteers was that the project was very rewarding and meaningful.

Boy Scout Stewardship Project at Refuge Gateway

In 2011, the DRIWR was approached by Boy Scout Troop # 802 of the Huron District within the Boy Scouts of America Great Lakes

Council about getting involved in a habitat restoration project in the Refuge. With the encouragement of the Troop Leader, local scout Nathan Lewis organized a shoreline restoration project for the outflow of the Monguagon wetland system at the Refuge Gateway in Trenton, Michigan. This restoration project fulfilled the requirements for Nathan to receive his Eagle Scout badge. Nathan's project included working with Refuge staff to scope out the project, help raise money, and recruit other scouts and parents to complete all restoration work. The total project cost was approximately $2,500, with some of the funds coming from the Boy Scouts and some coming from a local fishing club called the Downriver Walleye Federation.

This Boy Scout project entailed using live stakes and faschines to restore the shoreline of the outflow of the Monguagon wetland system at the Refuge Gateway. Live stakes are live wood cuttings that come from various native shrubs. Faschines are long bundles of live woody vegetation buried in a streambank in shallow trenches (Figure 30). Typically, live stakes are used to temporarily secure the fascines in trenches parallel to the flow of the stream. Together, live stakes and faschines help stabilize shorelines, prevent erosion, and enhance habitat. The shrubs eventually produce a canopy that provides habitat for birds and provides shade cover for the stream, promoting cooler water temperatures that provide greater oxygen for fish, amphibians, and insects.

The Monguagon wetland system is a restored wetland created through the process of daylighting (i.e. bringing it above ground and exposing it to daylight) a former underground storm water pipe. Historically, this underground pipe would transport storm water runoff from local impervious surfaces (roads and parking lots) directly into the Detroit River. Now, with the pipe removed, storm water flows through a settling pond and emergent wetland. The settling pond uses natural processes to settle out sediments and the emergent wetland uses native plants to remove nutrients. This Eagle Scout project was not only a "teachable conservation moment" for the scouts, but it will offer a unique opportunity to teach innovative storm water management at the Refuge Gateway to thousands of students who will come to the visitor center each year. It is another excellent example of citizen- and community-based stewardship.

Figure 30. Local Boy Scouts carrying a fascine to be installed along the outflow of the Monguagon wetland system at the Refuge Gateway in Trenton, Michigan (photo credit: Allison Krueger).

Citizen Stewardship at the Refuge Gateway

In the summer of 2012, the Refuge hired a Greening of Detroit work crew to assist with restoring coastal habitats at the Refuge Gateway and Humbug Marsh in Trenton and Gibraltar, Michigan, including planting trees in riparian and upland habitats, restoring wetlands, monitoring habitats, and constructing and maintaining trails. Four Greening of Detroit workers and a work crew leader were employed for six months during the spring and summer on this project. The Greening of Detroit workers had received hands-on training and completed Landscape Industry Technician certification (as defined by Michigan Green Industry Association).

During that summer, the Greening of Detroit Restoration Field Crew was responsible for restoring habitats at the Refuge Gateway as part of a larger effort to expand the ecological buffer of Humbug Marsh. In addition, as noted in Chapter 6, the Refuge Gateway was being restored as the future home of the Refuge's visitor center. The work crew was formed in a partnership between the Refuge and Greening of Detroit, a non-profit organization that works to improve ecosystem health in Detroit through tree planting projects, environmental education, urban agriculture, open space reclamation, and workforce development programs.

Leading the crew was Penelope Richardson-Bristol, a wetland ecologist from Eastern Michigan University. She came with a background in wetland restoration ecology, including research of soil seedbanks and vegetation monitoring on restoration sites. Her four-person crew was selected by the Greening of Detroit for their strong background in planting and caring for trees, and their Landscape Industry Technician certification.

The crew not only planted trees and assisted with other habitat restoration efforts, but they also spent a substantial amount of time assisting with community stewardship events, volunteer workdays, and open houses where they were able to share their knowledge and taught others (Figure 31). This crew literally helped build the capacity of the Refuge to restore habitats using citizen scientists and citizen stewards. Examples of specific stewardship tasks performed included:

- Installing live stakes that eventually would grow into small shrubs and trees;
- Assisting hundreds of volunteers to plant over 300 large native trees in 2012, many with over 136-kg (300-pound) rootballs;
- Watering new plantings, especially during the hot dry summer months, and installing tree anchors to promote strong root growth;
- Protecting newly planted trees and vegetation from deer browse and geese grazing; and
- Conducting plant surveys to monitor the effectiveness of restoration efforts.

This summer work effort was particularly successful because of the knowledge, dedication, and passion of both the Greening of Detroit work crew and the passion and commitment of citizens. Project oversight was provided by the Refuge's landscape designer – Allison Krueger. The combination of a landscape designer trained in landscape architecture, a wetland ecologist, the Greening of Detroit work crew, support from the IWRA, and passionate citizens resulted in both achieving substantial conservation results and providing a

Figure 31. Greening of Detroit work crew works with citizen stewards to plants trees at the Refuge Gateway, 2012 (photo credit: U.S. Fish and Wildlife Service).

compelling stewardship experience. Amazingly, more than 800 volunteers worked for more than 2,000 hours on these stewardship activities that spring and summer at the Refuge Gateway. Virtually all of them left with a sense of pride and ownership, and many ended up coming back to the Refuge for additional stewardship events in subsequent years.

Concluding Thoughts

People generally like to support the conservation of charismatic megafauna like Lake Sturgeon, Bald Eagles, and Peregrine Falcons. However, society generally underappreciates and often neglects non-charismatic components of the ecosystem, particularly in urban areas. For example, loss of wetlands, invasion of upland areas by exotic species, urban sprawl, and many other trends are dramatically altering the Detroit River ecosystem and negatively impacting its important ecosystem services. Clearly, more needs to be done to help reverse these trends and citizens can play a key role.

To recruit, educate, motivate, and inspire urban denizens to become conservationists requires a strong commitment to citizen science and stewardship. The DRIWR examples presented above are not comprehensive, but merely representative of the important work being carried out in this urban refuge.

Based on DRIWR experiences with citizen science and stewardship, the value and benefits of this work are substantial and include:

- developing a personal connection to the places citizens work and study;
- gaining an understanding of environmental and natural resource problems, challenges, and needs;
- learning about scientific methods and how science contributes to management;
- becoming involved in environmental and natural resource management decisions;
- building the capacity of governments, nongovernmental organizations, and other stakeholder groups to fulfill their environmental and natural resource missions; and
- improving scientific literacy and developing a stewardship ethic.

Achieving these benefits requires effective citizen science and stewardship project planning and implementation that:

- ensures measurable results,
- expands knowledge,
- provides meaningful experiences for volunteers, and
- ensures that volunteers have fun.

Indeed, Latimore and Steen (2014) have shown that partnerships between management agencies and local environmental organizations on volunteer monitoring and citizen science can provide long-term support for necessary research, monitoring, and management of aquatic ecosystems in Michigan.

Literature Cited

Conway, C. J., 2009. Standardized North American Marsh Bird Monitoring Protocols, version 2009-2. Wildlife Research Report #2009-02. U.S. Geological Survey, Arizona Cooperative Fish and Wildlife Research Unit, Tucson, Arizona, USA.

Craves, J., 2007. Detroit River Christmas Bird Count. In: J.H. Hartig, M.A. Zarull, J.J.H. Ciborowski, J.E.Gannon, E.Wilke, G. Norwood, A. Vincent (Eds.), *State of the Strait: Status and Trends of Key Indicators*, pp.243-246, Great Lake Institute for Environmental Research, Occasional Publication No. 5, University of Windsor, Ontario, Canada.

Latimore, J.A. Steen, P.J., 2014. Integrating freshwater science and local management through volunteer monitoring partnerships: the Michigan Clean Water Corps. Freshwater Science 33(2), 686-692.

Stein, J., 2011. Detroit River Hawk Watch. Hawk Migration Studies 38(1), 24-27.

Stein, J., Norwood, G., Conard, R., 2013. Detroit River Hawk Watch 2012 Season Summary. U.S. Fish and Wildlife Service, Detroit River International Wildlife Refuge, Grosse Ile, Michigan, USA.

United Nations World Commission on Environment and Development, 1987. *Our Common Future*. Oxford University Press, Oxford, United Kingdom.

U.S. Fish and Wildlife Service, 2005. Comprehensive Conservation Plan and Environmental Assessment for the Detroit River International Wildlife Refuge. Grosse Ile, Michigan, USA. www.fws.gov/midwest/planning/detroitriver/ (April 2013).

Yung, L., 2007. Citizen monitoring and restoration: Volunteers and community involvement in wilderness stewardship. U.S. Department of Agriculture, Forest Service Proceedings, Report No. RMRS-P-49, pp. 101-106, Fort Collins, Colorado, USA.

CHAPTER 9

Reconnecting People Back to the Land and Water Through Outdoor Recreation

The U.S. Fish and Wildlife Service (1999), in its vision document titled "Fulfilling the Promise," recognized that the heritage of the National Wildlife Refuge System is intertwined with the will of concerned citizens. By extension, that means that the heritage of the Detroit River International Wildlife Refuge (DRIWR) is intertwined with the will of concerned residents of southeast Michigan and southwest Ontario. Therefore, every effort must be made to educate and inform citizens about how they are part of the land/ecosystem, not separate from it, and about their natural resource inheritance and stewardship responsibility. Clearly, for the DRIWR and other urban refuges to reach their full potential will require better connecting with residents, educating them in the land/ecosystem ethic, and inspiring them to become stewards and to live sustainably.

One of the major challenges facing the DRIWR and other urban refuges is to become more relevant or in some cases to remain relevant to urban citizens who have many competing priorities and few outdoor natural resource experiences (USFWS, 2011). Clearly, it is a major challenge to find ways and means to connect with young people who are technologically fluent, but deficient in nature experiences. This chapter will share some of the things being done through the DRIWR to reconnect urban residents to the natural resources in their watershed through engagement in compelling outdoor recreational and educational experiences. The premise is that reconnecting people to the land and water through compelling outdoor recreational and educational experiences helps foster an appreciation and love for the outdoors. That, in turn, helps develop a strong sense of place that inspires positive actions, a sense of ownership, and stewardship for the community's natural capital.

Birding

No matter where you live, you can experience the joy of hearing and seeing birds. However, maybe to your surprise, the birding experience along the Detroit River and western Lake Erie is remarkable. The Detroit River and western Lake Erie are at the intersection of two major North American flyways – the Atlantic and Mississippi. Over 300,000 diving ducks, 75,000 shorebirds, and hundreds of thousands of landbirds and fall raptors frequent the shoreline habitats to rest, nest, and feed (Hartig et al., 2010). Over 30 species of waterfowl, 23 species of raptors, 31 species of shorebirds, and 160 species of songbirds are found along or migrate through this corridor (Hartig et al., 2010). This avian biodiversity and the diversity of habitats to support these birds have given the region international acclaim. The Detroit River and western Lake Erie have been recognized for their biodiversity in the North American Waterfowl Management Plan, the United Nations Convention on Biological Diversity, the Western Hemispheric Shorebird Reserve Network, the Biodiversity Investment Area Program of Environment Canada and U.S. Environmental Protection Agency, and in recent years as North America's only international wildlife refuge (Hartig et al., 2010).

Interest in bird watching is growing throughout the world, including along the Detroit River and western Lake Erie. The economic significance of this outdoor recreational pastime in 2006 was that 71.1 million wildlife watchers in the U.S. spent $45.7 billion on their wildlife watching activities around their homes and on trips away from their homes (U.S. Department of Interior, U.S. Fish and Wildlife Service and U.S. Department of Commerce, U.S. Census Bureau, 2006).

Over 300 species of birds have been identified in the Detroit River corridor by the Detroit Audubon Society. With knowledge of these many exceptional birding opportunities in the Detroit-Windsor metropolitan area, Metropolitan Affairs Coalition, the U.S. Fish and Wildlife Service, Michigan Sea Grant, the International Wildlife Refuge Alliance, Wild Birds Unlimited, and the National Fish and Wildlife Foundation developed a unique "Byways to Flyways" bird driving tour map to promote 27 exceptional birding sites throughout the Windsor-Detroit metropolitan area (Figure 32). Included within these sites are many Important Bird Areas (IBAs) identified by National Audubon Society, two "Wetlands of International Importance" identified under the international Ramsar Convention (i.e. Point Pelee National Park in Ontario and Humbug Marsh in Michigan), several Christmas Bird Count sites, and two internationally recognized hawk watch sites. These world-class birding opportunities are helping to reconnect many watershed residents to stopover habitats right in their backyard.

Fishing

Each spring nearly 10 million Walleye ascend the Detroit River from Lake Erie to spawn, attracting substantial numbers of anglers and international fishing tournaments. In fact, there are times when there are so many anglers that this phenomenon is called "migration madness." It should be no surprise to anyone that the Detroit River tournaments have landed numerous 5.9-6.4 kg (13-14 pounds) Walleye. It is both the numbers and size of Walleye that attract anglers from all over North America. During these major spring Walleye fishing tournaments so many fishing boats ply the Trenton Channel of the Detroit River that someone could almost literally cross the channel by hopping from boat to boat.

As one can imagine, angler interest in and excitement about Walleye fishing is tremendous. To capture, organize, and celebrate this interest, Downriver Walleye Federation was established to unite Walleye fishermen that ply the Detroit River and Lake Erie. The clubs primary goal is to exchange ideas and information about Walleye fishing in the "Walleye Capital of the World." Downriver Walleye Federation convenes monthly meetings that attract over 100 anglers and include engaging speakers. Throughout the year, members compete against each other in Walleye tournaments. Tournament fees are very low and competition is high (Figure 33). Club members have substantial experience in river jigging, hand line trolling, and Lake Erie deep water Walleye fishing tactics that they are willing to share with all. It should be no surprise that Walleye fishing on the Michigan side of the lower Detroit River alone brings in more than U.S. $1 million to local communities each spring.

As noted above, this exceptional Walleye fishery also attracts professional tournaments. FLW Outdoors has frequently held Walleye tournaments on the Detroit River offering over $500,000 in prize money. In 2007, U.S. Fish and Wildlife Service, Metropolitan Affairs Coalition, the Detroit Metro Sports Commission, Detroit Metro Convention & Visitors Bureau, and the City of Detroit Mayor's Office worked with FLW Outdoors to bring the "Chevy Open" bass fishing tournament to the Detroit River and western Lake Erie. Over 200 professional and 200 amateur anglers participated in a four-day bass tournament that offered $1.5 million in prize money. This "Chevy Open" also generated an additional $4-5 million in the economy related to lodging, fuel, food, equipment, etc.

Efforts are also being made to develop the next generation of anglers. We all recognize how most children love to fish. If you have ever seen the excitement in a child's face when she or he catches a fish, it literally brings a smile to your face. Fishing is one of the ways

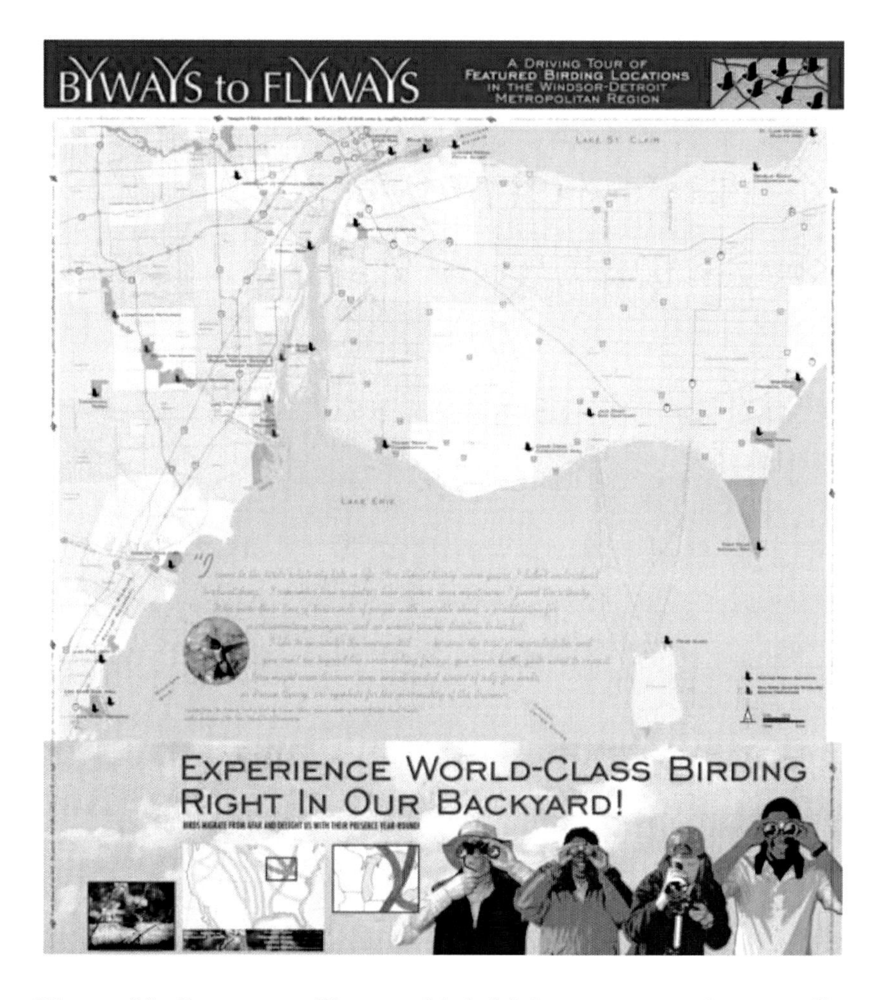

Figure 32. Byways to Flyways bird driving tour map (map credit: Metropolitan Affairs Coaliton).

Figure 33. Trophy Walleye (6 kg or 13.5 pounds) caught in a Downriver Walleye Federation Walleye tournament, 2006 (photo credit: Downriver Walleye Federation).

to help instill a sense of wonder for natural resources and discover the joy, excitement, and mystery of the world children live in.

The Detroit River Days' Kids Fishing Fest is a youth fishing and conservation program launched by Detroit Riverfront Conservancy, U.S. Fish and Wildlife Service, Michigan Department of Natural Resources, and many other partners in 2011. It has quickly become an annual tradition. The Detroit River Days' Kids Fishing Fest is designed to engage children ages 6 to 14 and their families in the sport of fishing along the Detroit River and to educate them on the principles of water stewardship, while creating lasting memories and strengthening ties to their families, communities, and the environment.

Detroit River Days' Kids Fishing Fest is held in downtown Detroit at Milliken State Park along the Detroit RiverWalk. In 2012, more than 500 children and their families were treated to a day of free fishing and family fun, including fishing seminars, food, entertainment, arts and crafts, face painting, prizes, and a raffle drawing (Figure 34).

Figure 34. Kids Fishing Fest is held in downtown Detroit at Milliken State Park along the Detroit RiverWalk, 2012 (photo credit: Detroit Riverfront Conservancy).

This Kids' Fishing Fest serves as a kick off to Detroit River Days festivities that attract hundreds of thousands people to the waterfront over a three-day period. The U.S. Fish and Wildlife Service has provided $5,000 in grant funding under Connecting People With Nature for purchasing fishing equipment and other materials for Kids Fishing Fest. These efforts have been very successful in generating excitement for youth fishing in inner city and for developing lasting memories with family members and friends.

Hunting

As waterfowl migrate from the Arctic Circle and Canada to overwintering grounds in the Caribbean and Gulf of Mexico they need places to stop, rest, and feed. These waterfowl, often in tens to hundreds of thousands, come to staging and feeding areas along the

Detroit River and western Lake Erie, attracted by rich beds of Wild Celery and other food. It often seems paradoxical to see this number of waterfowl staging and feeding in the shadow of big industry along the Detroit River. It is even more amazing to see them lift off the water in such density that they sometimes blacken the sky. These transcontinental waterfowl migrations are clearly one of nature's great spectacles.

The National Wildlife Refuge System Improvement Act of 1997 defined hunting as a legitimate and appropriate general public use of the National Wildlife Refuge System (U.S. Fish and Wildlife Service, 1999). Further, the Comprehensive Conservation Plan for the DRIWR states that one of the goals is to facilitate and promote hunting, fishing, wildlife observation, wildlife photography, environmental education, and interpretation as wildlife-dependent uses (U.S. Fish and Wildlife Service, 2005).

Consistent with the National Wildlife Refuge System Improvement Act of 1997 and as called for the Refuge's Comprehensive Conservation Plan, a Hunt Plan was completed in 2011 with considerable public input. Following approval of the U.S. Fish and Wildlife Service, nearly 101 ha (250 acres) of Refuge lands were opened to hunting in 2012. Humbug, Calf, and Sugar Islands, as well as the Strong and Fix Units, were opened to small game, big game and migratory bird hunting in accordance with state regulations. Coastal portions of the Plum Creek Bay Unit, which are only accessible by water, and the Brancheau Unit were both opened for migratory bird hunting only. The refuge was able to provide hunting at the Brancheau Unit through an effective partnership with the Michigan Department of Natural Resources at the Pointe Mouillee State Game Area. Hunting was coordinated with the daily drawing that occurs as part of the State Game Area's managed waterfowl hunt.

According to reports from citizens and law enforcement officials, all 2012 hunts were well received and quite successful. At the Brancheau Unit waterfowl hunt, there were only two days out of the regular waterfowl season that none of the zones were selected in the daily drawing. A total of 103 waterfowl hunters were able to hunt the Brancheau Unit alone during the 2012 season and the total duck harvest for the 15 days that the unit was hunted was 75 ducks (Figure 35). Green-winged Teal, Wood Duck, and Mallard were the top three species harvested. It should be noted that Michigan Department of Natural Resources was instrumental in making the Brancheau Unit hunt a success. This partnership with the Michigan Department of Natural Resources resulted in increasing hunting opportunities in southeast Michigan and enhancing the reputation of Detroit being one of the top ten metropolitan areas for waterfowl hunting in the United States (Card, 2013). It should also be noted that the U.S. Fish and Wildlife Service annually supports the Gibraltar Duck Hunter's Association in its youth duck hunt that exposes over 50 young boys and girls to waterfowl hunting in an effort to promote and sustain the hunting tradition in southeast Michigan.

Figure 35. Waterfowl hunting at the Refuge's Bracheau Unit on opening day, 2012 (photo credit: Michigan Department of Natural Resources).

Paddling

When most people think about kayaking and canoeing, they think about paddling through pristine rivers or lakes, or through streams running through pristine forests. However, paddling enthusiasts in metropolitan Detroit are saying think again.

In 2006, Metropolitan Affairs Coalition, a nonprofit organization made up of business, government, labor, and educational leaders dedicated to building consensus and seeking solutions to regional issues in metropolitan Detroit, developed a vision plan for a Detroit

Heritage River Water Trail. It is a river version of a greenway trail (or "blueway") that provides opportunities for canoeing, kayaking, and small boat paddling. Metropolitan Affairs Coalition brought together, under its Greater Detroit American Heritage River Initiative, numerous partners, including the Downriver Linked Greenways Initiative, Friends of the Detroit River, Huron-Clinton Metropolitan Authority, Michigan Sea Grant, National Park Service, the U.S. Fish and Wildlife Service, and Wayne County Parks. These partners then developed a vision plan for this unique water trail that would both promote existing close-to-home paddle-based recreational opportunities and to plan for future expansion of ecotourism. Research has shown that interest in paddle-based recreation was increasing and that this recreation industry contributes $36.1 billion annually to the U.S. economy (Outdoor Industry Foundation, 2006).

What makes this effort so unique is that not only can it provide paddlers with an exceptional paddling experience through an international wildlife refuge and around 23 islands and numerous shoals, but it can provide an unparalleled paddling experience through the industrial heartland and among freighters and skyscrapers. In 2013, the vision plan was updated to a usable water trail map that identifies both paddling access points and paddling routes (Figure 36). Paddling opportunities extend some 61.2 km (38 miles) from Maheras Gentry Park at the upper end of the Detroit River to Luna Pier on western Lake Erie in the southern end of Monroe County. Kayakers can paddle through time along waterways that sustained Native Americans, supported the Fur Trade, fostered the Industrial Revolution, and are now heralded as North America's only international wildlife refuge. Along the way, paddlers can see the 396-ha (980-acre) island park designed by Frederick Law Olmstead (i.e. Belle Isle), the oldest rowing club (i.e. Detroit Rowing Club) in the United States established in 1839, the international crossing of the Underground Railroad, the historical drydocks that were used to build 650 freighters during the ship building era, the industries that made Detroit the "Arsenal of Democracy," and truly exceptional wildlife, including waterfowl, Egrets, Bald Eagles, Herons, and countless fish being landed by anglers. Such paddling experiences are reconnecting people to the rich biodiversity and exceptional natural resources in their backyard.

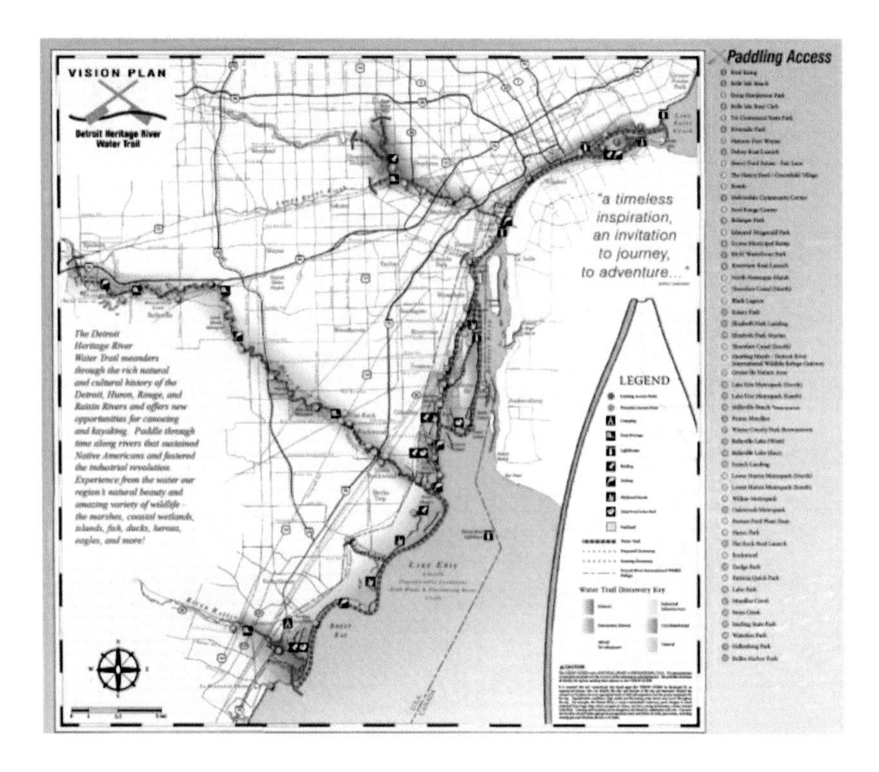

Figure 36. Detroit Heritage River Water Trail along the Detroit River and western Lake Erie (credit: Metropolitan Affairs Coalition).

As part of the Detroit Heritage River Water Trail, a kayak landing was constructed at the Refuge Gateway in 2011. Such amenities not only enhance ecotourism in the area, but substantially improve interpretation and environmental education opportunities in the effort to help develop the next generation of conservationists and sustainability entrepreneurs.

Researchers from Michigan State University agree and have concluded that the Detroit River is now a draw for paddling enthusiasts. Dr. Christine Vogt of Michigan State University's Department of Community, Agriculture, Recreation and Resource Studies has noted:

The leadership and collaboration toward water trails and paddling sports in the Downriver area is terrific and forward-thinking. This recreation and tourism activity can put Michigan on the map with all our Great Lakes and rivers in urban, suburban and rural areas.

Greenways

Greenways are linear open spaces, including habitats and trails, that link parks, nature areas, cultural features, or historic sites with each other, for recreation and conservation purposes. Based on experience throughout North America, greenways promote outdoor recreation, catalyze economic development, increase adjacent property values, celebrate historical and cultural assets, promote conservation and environmental education, and improve quality of life. Greenways can provide an exceptional outdoor recreational experience that reconnects children and families to natural resources, and that builds a stewardship ethic. It should not be surprising that greenways are an enormous source of community pride.

In many respects, southwest Ontario's greenway trails have been an inspiration to the development of southeast Michigan's greenway trails. Windsor's Department of Parks and Recreation maintains 12 km^2 (3,000 acres) of green space, 180 parks, and 64 km (40 miles) of trails. Beginning in the 1960s, Windsor started creating a shared-usage trail network called the "Windsor Loop" that circumnavigates around the entire city and connects to neighboring communities. The longest greenway trail in this network is the Roy A. Battagello River Walk (built in the late-1960s, and upgraded/widened several times), stretching from west of the Ambassador Bridge to the historical Hiram Walker Distillery, a distance of about 8 km (five miles).

This trail also connects to other trails leading to key natural areas and city parks, including Ojibway Park and Ojibway Prairie Provincial Nature Reserve, Malden Park, Spring Garden Area of Natural Scientific Interest, and others. Other key extensions of this greenway trail network include the Ganatchio Trail (built in 1971), its Little River Extension (built in 1996), and the Devonwood Bike Trail (built between the mid-1980s and the early-1990s).

The City of Windsor greenways also link to a 50-km regional greeenway trail called the Chrysler Canada Greenway, one of Canada's most beautiful trails that connect 25 otherwise separate natural areas in Essex County. Along the Chrysler Canada Greenway you can see unique natural resources, rich agricultural lands, historically and architecturally significant structures, and award winning wineries.

In southeast Michigan, greenways have been championed by the Community Foundation for Southeast Michigan, the Downriver Linked Greenways Initiative, the Greater Detroit American Heritage River Initiative, the Michigan Rails to Trails Conservancy, and numerous communities. Although considerable planning was done for nearly three decades, greenway construction and development did not occur substantially until the late-1990s. Considerable progress has been made in the last 10-15 years, including:

- A $25 million investment in greenways by the Community Foundation for Southeast Michigan that has leveraged an additional $90 million to construct over 160 km (100 miles) greenways in southeast Michigan;
- Downriver Linked Greenways Initiative has championed the construction of over 80 km (50 miles) of continuous greenways in the watershed of the lower Detroit River; and
- Detroit Riverfront Conservancy has raised over $90 million in the last 10 years to construct the Detroit RiverWalk in Downtown Detroit, representing one of the largest urban waterfront development projects in the United States.

The combined long-term vision for greenways is to achieve an interconnected trail system that stretches from Lake Huron to Lake Erie, up the tributaries like the Clinton and Rouge rivers, Ecorse Creek, and the River Raisin, and across to Canada. The Refuge Gateway, the future home of the Refuge's visitor center on the U.S. side, is now connected with 80 km (50 miles) of continuous greenway trails and thereby connected with numerous communities. Indeed, these greenways have already made a huge impact in improving awareness of unique natural resources and are laying the foundation to foster a stewardship ethic.

Signature Events That Engage People in Conservation and Outdoor Recreation

Another way of engaging urban residents in compelling and fun conservation experiences is through signature and special events. These events are held to promote awareness, to generate enthusiasm for the outdoors, and to be fun. Selected examples of such events include:

- International Migratory Bird Day recognizes the movement of nearly 350 species of migratory birds from their wintering grounds in South and Central America, Mexico, and the Caribbean to nesting habitats in North America, and is celebrated on the second Saturday in May at three locations in Canada's Priority Natural Area and two DRIWR locations;
- World Wetlands Day is observed worldwide each year on February 2nd to celebrate in importance of wetlands, including "Wetlands of International Importance" designated under the Ramsar Convention in 1971 (over 1,200 high school students annually participate in a wetlands' program and exposition at Gibraltar Carlson High School);
- National Wildlife Refuge Week is celebrated each October to showcase the premier land conservation network in the world that includes DRIWR;
- Pointe Mouillee Waterfowl Festival is held each year at Point Mouillee State Game Area on the first weekend after Labor Day and right before duck hunting season opens – it is an annual tradition that showcases duck hunters' abilities, celebrates wildlife art and hunting equipment, and attracts 8,000-10,000 enthusiasts;
- U.S. Fish and Wildlife Service and its Friends' Organization called the International Wildlife Refuge Alliance are partners in the annual Hawk Fest sponsored by Huron Clinton Metropolitan Authority at Lake Erie Metropark that is considered one of the three best places to watch hawk migrations in the United States – annual attendance is over 4,000;
- The annual Detroit River Days is premier waterfront event on the Detroit RiverWalk that celebrates maritime history, ecological revival, culinary culture, and family-friendly activities and entertainment – it attracts hundreds of thousands of people; and
- Eagle tours have become an annual tradition where participants can watch over 100 Bald Eagles that overwinter at DTE Energy's Monroe Power Plant.

Such events are an excellent way to engage citizens in exciting outdoor activities and offer a unique opportunity to educate them about how conservation is vital to their lives.

Concluding Remarks

Although nature may seem far away to most urban residents and conservation agencies and organizations often struggle to explain its benefits, most people agree that nature is vital to our health and well-being, and helps nourish our sense of wonder, imagination, and curiosity (USFWS, 2011). Simply put, we need to reconnect people with nature as part of a long-term strategy to inspire individual respect, love, and stewardship of the land to be able to develop a societal land/ecosystem ethic that achieves sustainability. This will help establish conservation relevance in a growing urban population. And it will help foster a more informed citizenry that actively supports and understands the value of conservation.

Just like a carpenter needs many tools to build a house, reconnecting people back to the land and water to help develop and land/ecosystem ethic will require many tools and techniques. Some tools, like the ones identified earlier in the chapter, are traditional and well-tested, but others are new and untested. We can't be afraid to experiment with new tools and techniques to reconnect urban residents with nature.

Louv (2005) has shown in his book titled *Last Child in the Woods* that today children's physical contact and intimacy with nature is fading. Louv (2005) directly links the lack of nature in the lives of today's generation (i.e. nature deficit) to some of the most disturbing childhood trends, such as the rise in obesity rate, attention disorders, and depression. This hurts our children, our families, our communities, and our environment. Luckily for us, the cure starts in our own backyards and can be as simple as building a bat house or a backyard weather station, planting a garden or native landscaping, collecting lightning bugs at dusk or worms at night, building a tree house, raising caterpillars for eggs, taking a nature hike, visiting a local conservation area, state park, or wildlife refuge, or numerous other outdoor activities.

Urban wildlife refuges and other urban conservation areas, because of their proximity to so many people, can play a truly unique and important role in addressing nature deficit. Urban refuges have the unique proximal natural resources to help children experience nature as the supporting fabric of their everyday lives. Whether it's hiking, fishing, hunting, birding, learning through environmental education, photography, environmental interpretation, or just plain exploring in the woods, urban refuges have what educators, city planners, business leaders, and parents want – unique natural resources that can enhance quality of life, contribute to ecosystem health and healthful living, and nourish our sense of wonder, imagination, and curiosity.

If a city planner or a landscape architect were designing a future green city that would eliminate nature deficit, they would undoubtedly want an urban wildlife refuge or other conservation area. From a design perspective it is a unique and compelling asset to have an urban refuge that offers exceptional wildlife-dependent outdoor recreation, enhances quality of life, increases community pride, and gives competitive advantage. Urban refuges and urban conservation areas provide unique opportunities for people to become involved in a positive way with the natural world and to have a personal stake in the stewardship of these natural resources.

Think about it, urban conservation areas are very special places that offer urban denizens the opportunity to sit and listen to a morning chorus of birds or to the rhythmic rumbling of frogs in a wetland. In places like DRIWR, they also offer urbanites the opportunity to watch awe-inspiring waterfowl migrations, including extraordinary kettles of hawks wheeling and circling in the air. Urban refuges and other urban conservation areas not only give city dwellers brief escapes and moments of peace in an otherwise busy and stressful day, but they provide high quality recreational opportunities like fishing, hunting, wildlife observation, wildlife photography, environmental education, and environmental interpretation.

Again, it is critically important that government agencies, educational institutions, businesses, environmental organizations, conservation clubs, faith-based organizations, and concerned citizens join forces to help reconnect people to the land and water in urban areas through compelling outdoor recreational and educational experiences that help foster an appreciation and love for the outdoors. That, in turn, will help develop a strong sense of: place that inspires positive actions; ownership; and stewardship for the community's natural resources. Finally, each of us must take personal responsibility to provide opportunities for youth to experience the natural world so that they can learn to love, respect, and care for it, and hopefully someday do the same for their children.

Literature Cited

Card, J., 2013. Ten great cities for waterfowlers. Ducks Unlimited, Inc. Memphis, Tennessee, USA. http://www.ducks.org/hunting/destinations/10-great-cities-for-waterfowlers (March 2013).

Hartig J.H., Robinson, R.S., Zarull, M.A., 2010. Designing a Sustainable Future through Creation of North America's only International Wildlife Refuge. Sustainability 2(9), 3110-3128. www.mdpi.com/2071-1050/2/9/3110/pdf (April 2013).

Louv, R., 2005. *Last Child in the Woods: Saving Our Children from Nature-Deficit Disorder.* Algonquin Books of Chapel Hill, Chapel Hill, North Carolina, USA.

U.S. Department of the Interior, U.S. Fish and Wildlife Service, U.S. Department of Commerce, U.S. Census Bureau, 2006. National Survey of Fishing, Hunting, and Wildlife-Associated Recreation. Washington, D.C., USA.

U.S. Fish and Wildlife Service (USFWS), 1999. *Fulfilling the Promise: Visions for Wildlife, Habitat, People, and Leadership.* National Wildlife Refuge System, Arlington, Virginia, USA.

USFWS, 2005. *Comprehensive Conservation Plan and Environmental Assessment for the Detroit River International Wildlife Refuge.* Grosse Ile, Michigan, USA. www.fws.gov/midwest/planning/detroitriver/ (April 2013).

USFWS, 2011. *Conserving the Future: Wildlife Refuges and the Next Generation.* National Wildlife Refuge System, Arlington, Virginia, USA.

CHAPTER 10

Lessons Learned

Frederick Law Olmsted (1822-1903) was an American journalist, social critic, a public administrator, and famous landscape designer. Today, he is popularly considered the father of American landscape architecture. Most people know Olmstead through the physical legacy of stunning landscapes, including: Central Park in New York City; Belle Isle in Detroit; the United States' first and oldest coordinated system of public parks and parkways in Buffalo, New York; the United States' oldest state park called the Niagara Reservation in Niagara Falls, New York; Mount Royal Park in Montreal, Quebec; the Emerald Necklace in Boston, Massachusetts; Highland Park in Rochester, New York; the Grand Necklace of Parks in Milwaukee, Wisconsin; the George Washington Vanderbilt II Biltmore Estate in Asheville, North Carolina; Montebello Park in St. Catherines, Ontario; Jackson Park in Chicago, Illinois; and many other well-known urban parks and urban natural areas (Rybczynski, 1999).

Olmsted believed that the crowded urban environment was unhealthy and that native urban landscapes strengthened society by providing a place where all classes could mingle in contemplation and enjoyment of nature (Rybczynski, 1999). He believed that the perfect antidote to the stress and artificialness of urban life was a nice stroll through the natural beauty of a pastoral park or native landscape. He literally saw urban parks and conservation areas as places of harmony, where people would go to escape urban life and reconnect with nature. Olmstead also wanted these urban parks to be as close to as many people as possible and to be available to all people from all walks of life.

Olmstead's genius was how he used naturally occurring features and native landscapes to give opportunities to relax in the outdoors and to increase cultural and recreational opportunities in cities. His urban designs also seek to advance interaction and communication among urban residents, leading to a sense of shared community and stewardship. He saw his role as helping shape the American city by designing public parks and park systems that would reconnect people with nature and reap their recreational and quality of life benefits. One could easily call Olmstead one of the first urban conservationists. Indeed, thanks to Olmsted's vision and creative genius, many large American cities have a little bit of nature left in them.

Today, Olmstead's design principles, that emphasized natural landscapes in cities that were accessible to all, are very important foundational building blocks to help natural resource agencies deliver conservation in major urban areas and to help develop the next generation of conservationists. Clearly, the conservation community needs to become more actively involved in a cooperative effort that delivers conservation in urban areas and that makes nature experiences part of everyday urban life (Kareiva, 2013). Most conservationists avoid cities and want to work in pristine or wilderness areas. Furthermore, when scientific assessments are made, most urban areas are found to be too degraded to rank high enough on conservation priority lists (Kareiva, 2013). Candidly, this must change and much greater effort must be placed on urban conservation. The conservation community, the urban planning community, the landscape architecture community, the urban recreation community, the health care community, the educational community, faith-based organizations, the business community, nongovernmental organizations, and other key stakeholder groups need to come together in a cooperative conservation effort focused on the unique natural resource assets of the urban area, including city parks, metroparks, urban state parks or refuges (if present), greenway trail systems, or other urban conservation areas, or any network of these lands (Table 13).

One of the key questions is: "Who should lead this urban conservation initiative?" Should each urban area have its own unique local leadership? Should one governmental agency or nonprofit organization provide key leadership on a comprehensive urban conservation framework and work with local organizations to deliver it? Should a coalition of conservation agencies and organizations form a partnership to catalyze and guide an urban conservation initiative? The key point is that there is a dearth of conservation leadership in most urban areas. Further, some organizations and people are concerned that the level of effort and amount or resources required to deliver an effective urban conservation initiative are just too great and would dilute conservation of pristine/wilderness areas and other areas ranked higher on a conservation priority list. I tend to think that a coalition of agencies and organizations should form a partnership to catalyze and guide urban conservation initiatives in a manner that promotes local ownership and cooperative learning, and shares information on best practices and associated benefits.

Table 13. Selected examples of urban conservation and outdoor recreational initiatives that are helping make nature experiences part of everyday urban life as part of a long-term strategy to inspire individual respect, love, and stewardship of the land/ecosystem.

Name	Location	Description of Urban Conservation Initiative
Chicago Wilderness	Chicago, Illinois	More than 260 public and private partners are involved in restoring, protecting, and connecting 728,400 ha (1.8 million acres) through conservation and thoughtful, sustainable development practices
Olmstead Parks and Parkways System	Buffalo, New York	Buffalo's Olmstead Parks and Parkways System is the United States' oldest coordinated system of outdoor recreational spaces, designed by Frederick Law Olmsted and Calvert Vaux between 1868 and 1896. This urban park system consists of 485 ha (1,200 acres) of beautifully designed parks, parkways, and circles which weave throughout the City of Buffalo. It is on the National Register of Historic Places and is maintained by the Buffalo Olmsted Parks Conservancy. More than one million people use Buffalo's Olmsted Park System annually for recreation, relaxation, and rejuvenation.
Willamette Valley	Portland, Oregon	Audubon Society of Portland, Oregon Department of Fish and Wildlife, U.S. Fish and Wildlife Service, and other partners are collaborating on conservation of more than 4,046 ha (10,000 acres) of parks and natural areas that will also ensure access to nature for all residents.
The Los Angeles River Urban Wildlife Refuge Vision	Los Angeles, California	The Santa Monica Mountains Conservancy, a landscape-scale state agency, has helped preserve more than 24,281 ha (60,000 acres) of parkland and improve more than 114 public facilities in Southern California, with the goal of creating "an interlinking system of urban, rural and river parks, open space, trails, and wildlife habitats that are easily accessible to the general public." Their vision is to create an urban wildlife refuge centered on the long-mistreated Los Angeles River that may one day be entered into the National Wildlife Refuge System.

Table 13 Cont'd

Waterfront Regeneration Trust	Toronto, Ontario, Canada	The Trust helped rediscover the value and attractiveness of the water's edge in Canada's largest city with over five million people. In 1995, the Trust opened the Waterfront Trail, a 350-km (217-mile), virtually continuous trail along the Lake Ontario shoreline, which connects hundreds of parks, historic and cultural sites, wildlife habitats and recreation areas.
Wetlands Park	Las Vegas, Nevada	Las Vegas draws nearly 38 million annual visitors. Wetlands Park is a 1,173-ha (2,900-acre) wildlife refuge located in close proximity to the Las Vegas Strip. The primary purpose of the park is to reduce the environmental impact of the wastewater and storm water runoff leaving the drainage basin by building constructed wetlands. This project used water control structures, dams, and weirs to slow down the flow of water, catch silt, reduce the undercutting of dirt walls that form the Las Vegas Wash, and create unique wildlife habitats. The secondary purpose of the park is to promote environmental education and close-to-home outdoor recreation.
Stanley Park	Vancouver, British Columbia, Canada	Stanley Park is a world renowned 405 ha (1,000-acre) urban park and tourist attraction that attracts eight million annual visitors. It is a magnificent green oasis in the midst of the heavily built urban landscape of Vancouver. Stanley Park has been designated a national historic site in Canada and offers a wide range of nature experiences for all ages and interests. Urban residents can explore West Coast rainforest habitats, enjoy scenic views of water and mountains, and observe wildlife as part of an urban outdoor recreational experience. The famous a 22-km (14-mile) Seawall Path attracts 2.5 million annual pedestrians, cyclists, and joggers.
San Francisco Bay National Wildlife Refuge Complex	San Francisco, California	This Complex is a collection of seven National Wildlife Refuges (NWR): Antioch Dunes NWR; Don Edwards San Francisco Bay NWR; Ellicott Slough NWR; Farallon NWR; Marin Islands NWR; Salinas River NWR; and San Pablo Bay NWR. These wildlife refuges are devoted to protecting wildlife in unique habitats such as sand dunes, salt marshes, rocky offshore islands, mudflats, and uplands. Despite booming industries and growing populations, these National Wildlife Refuges preserve an incredibly complex ecosystem.

Table 13. Cont'd

Emerald Necklace	Boston and Brookline, Massachusetts	The Emerald Necklace consists of a 445-ha (1,100-acre) chain of parks linked by parkways and waterways in Boston and Brookline, Massachusetts. This linear system of parks and parkways was designed by Frederick Law Olmsted to connect the Boston Common. The Emerald Necklace Conservancy was created to protect, restore, maintain, and promote the landscape, waterways, and parkways of the Emerald Necklace park system as special places for people to visit and enjoy.
Grand Necklace of Parks	Milwaukee, Wisconsin	Milwaukee County Park System features over 140 parks and parkways totaling nearly 6,070 ha (15,000 acres) that offer a source of recreational enjoyment for citizens and visitors alike. Milwaukee's "crown jewel" is the Grand Necklace of Parks designed by Frederick Law Olmstead.
Detroit River International Wildlife Refuge	Detroit Metropolitan Area, Michigan	This refuge is the only international one in North America and one of the few truly urban ones in the U.S. Unique public-private partnerships are building the refuge and helping provide a compelling conservation and outdoor recreational experience to nearly seven million people in a 45-minute drive.
Multiple Habitat Conservation Program	San Diego County, California	The Multiple Habitat Conservation Program (MHCP) is a comprehensive conservation planning process that addresses the needs of multiple plant and animal species in San Diego County. It is made up of three subregional habitat planning efforts that contribute to preservation of regional biodiversity through coordination with other habitat conservation planning efforts throughout southern California. As an example, the MHCP for the cities of Carlsbad, Encinitas, Escondido, Oceanside, San Marcos, Solana Beach, and Vista has a goal to conserve approximately 7,689 ha (19,000) acres of habitat, of which roughly 3,561 ha (8,800 acres) (46%) are already in public ownership and contribute toward the habitat preserve system for the protection of more than 80 rare, threatened, or endangered species.
John Heinz National Wildlife Refuge at Tinicum	Philadelphia, Pennsylvania	This refuge is a green respite nestled within the urban setting of the city of Philadelphia. Refuge lands are a thriving sanctuary teeming with a rich diversity of fish, wildlife, and plants native to the Delaware Estuary. With partners' support, the refuge is both a leader in freshwater tidal marsh conservation and an ambassador of bringing conservation to an urban area with over 35 million Americans living within a two-hour drive.

Based on the evidence presented in the previous nine chapters, the Detroit River International Wildlife Refuge (DRIWR) can be viewed as a successful experiment in conserving continentally-significant fish and wildlife in a major urban industrial area, and in helping make nature experiences part of everyday urban life. However, there are many ways and means of fostering conservation in urban areas.

One of the goals of this book is to share lessons learned from this experiment of developing and building the DRIWR. Presented below are ten lessons learned about how this urban wildlife refuge has, thus far, been molded, fashioned, and formed in an effort to conserve continentally-significant wildlife, while making nature experiences part of everyday urban life to help develop a conservation ethic. All lessons may not be applicable to all urban areas and no priority ranking is intended in the numbering scheme. It is hoped that these lessons will help encourage more urban conservation initiatives throughout the world.

LESSON 1: Establish a Compelling Vision

Establishing a clear and compelling vision is a critical step in any major urban initiative, particularly when the number of stakeholders is large. It cannot be ambiguous. For example, contrast the vision of "We will be a leader in space exploration" with "We will put a man on the moon in 10 years." The latter has much more objective clarity and is much more compelling.

A vision must also be relevant, appealing, and engaging, and must be a picture that all stakeholders can carry in their hearts and minds (Senge 1990). Clearly, the Conservation Vision for the Lower Detroit River Ecosystem (Metropolitan Affairs Coalition, 2001) was effective because it was clear, specific, and strong. The vision was to create an international wildlife refuge, yet it was flexible enough to allow two countries and numerous agencies and other stakeholder groups to operationalize it within their unique institutional frameworks. Since the establishment of the DRIWR in 2001, over 300 public and private partners have made significant contributions to building or improving the refuge. Some examples of DRIWR accomplishments achieved through innovative public-private partnerships are presented below:

- Agencies have documented and celebrated the return of Lake Sturgeon, Lake Whitefish, Walleye, Bald Eagles, Peregrine

Falcons, Osprey, Wild Celery, Mayflies, Beaver, representing one of the most dramatic ecological recoveries in North America;
- A ByWay to FlyWays Bird Driving Tour Map highlights 27 exceptional birding locations in southeast Michigan and southwest Ontario;
- 53 soft shoreline engineering projects have been undertaken in the watershed of the Detroit River and western Lake Erie;
- International Migratory Bird Day is celebrated annually in five locations in Canada and the United States each May;
- The Fighting Island Sturgeon reef was constructed and successful with natural reproduction, representing the first ever Canada-U.S. funded fish habitat restoration project in the Great Lakes;
- Canadian and United States' partners have reached agreement on a common tern restoration target for the corridor and in 2012, after habitat restoration on Belle Isle, recorded two common terns fledging there for the first time since the 1960s;
- Biennial Canada-U.S. State of the Strait Conferences have been held since 1998 to assess ecosystem status and provide advice to improve research, monitoring, and management; and
- Citizen science is used to systematically monitor raptor migrations across the Detroit River at sister locations of Holiday Beach in Amherstburg, Ontario, Canada and at Lake Erie Metropark in Brownstown, Michigan, USA – representing one of the three best places to watch Hawk migrations in North America.

With the trend of most urban residents losing their connection with nature, there is an urgent and compelling need to reconnect them with nature as part of a long-term strategy to inspire individual respect, love, and stewardship of the land. One thing that could be done in any urban refuge or an urban conservation initiative is to make sure that its vision statement focuses on both wildlife conservation and connecting children and families with nature.

One particularly good example of a major urban area with a compelling vision that addresses both natural resource conservation and providing meaningful nature experiences to children and families is Chicago Wilderness. Chicago Wilderness is an alliance of more than 260 public and private organizations that work together to understand, protect, restore, and better manage the natural ecosystems in the Greater Chicago region that stretches from southwest Michigan, through northwest Indiana and northeast Illinois, and through southeast Wisconsin.

Chicago Wilderness' Green Infrastructure Vision is to restore, protect, and connect 728,400 ha (1.8 million acres) through conservation and thoughtful, sustainable development practices (Chicago Wilderness, 2004). Clearly, the Greater Chicago region has been blessed with some of the best remaining remnants of tall grass prairies, oak savannas, and wetlands, and many other rare plants and animals, including nearly 200 species listed as endangered or threatened in Illinois (Shore, 1997). And yes, this makes the Greater Chicago region one of only a handful of metropolitan areas worldwide that has such a high density of globally significant ecosystems. As a result, Chicago Wilderness makes a unique contribution to the conservation of global biodiversity and simultaneously helps enrich quality of life for local residents.

Chicago Wilderness promotes the integration of nature and wildlife into the places where people live and makes nature experiences part of everyday life. That's what makes Chicago Wilderness so unique. The work of the Chicago Wilderness consortium is coordinated by an Executive Council made up of representatives from the more than 260 public and private partner organizations. This unique consortium "lifts up" and celebrates the natural capital that helps make Chicago unique – locally, regionally, nationally, and globally. In essence, the Chicago Wilderness educates residents about unique local ecosystems, motivates and inspires them to protect, restore, and enhance these ecosystems, and manifests and celebrates how their ecosystems and the ecosystem services they provide improve quality of life. It is truly a unique approach that builds community pride, furthers conservation and sustainability, and attracts champions. Chicago Wilderness is now taking action through four key initiatives to achieve its Green Infrastructure Vision (Table 14).

Chicago Wilderness is a unique blend of effective education, innovative advocacy, creative marketing, and grassroots action. Further, it targets children to creatively and effectively reconnect them with nature. Clearly, this innovative work of the Chicago Wilderness is a model for other major urban areas interested in furthering urban conservation, protecting biodiversity, fostering sustainable development, and meaningfully involving families and children.

Table 14. Chicago Wilderness' four key initiatives to achieve its Green Infrastructure Vision (Chicago Wilderness, 2004).

Chicago Wilderness Initiative	Description
Restoring Nature	Members of Chicago Wilderness are making significant improvements in the ecological health of the region's natural systems and engaging residents as stewards of the region's lands and waters
Climate Action	Member organizations are developing strategies to mitigate climate change through land and water conservation action, and implementing the Chicago Wilderness Climate Action Plan for Nature, the first plan of its kind to link climate change specifically to issues of biodiversity conservation
Leave No Child Inside	Member organizations are raising public awareness about the importance of access to nature for healthy childhood development and providing the places and programs for generations of families to connect with nature
Green Infrastructure	Member organizations are creating a region where healthy ecosystems contribute to economic vitality and a high quality of life for all residents

Recommendation: It is recommended that urban stakeholder groups reach agreement on a compelling vision that focuses both on conserving urban wildlife and requisite habitats, and on making nature part of everyday urban life through inspiring conservation education, citizen science, stewardship, and outdoor recreational activities. This vision should be focused on the unique natural resource assets of the urban area, including city parks, metroparks, urban state parks or refuges (if present), greenway trail systems, or other urban conservation areas, or any network of these lands.

LESSON 2: Use Sound Science, Practice Adaptive Management, and Be Prepared to Compromise

Science is the objective and verifiable search for the truth. Sound science is a popular term being used to describe a systematic effort by qualified individuals of performing experiments, conducting investigations, and undertaking monitoring and surveys that lead to verifiable results and conclusions. It should go without saying that urban conservation initiatives need to use sound science.

The U.S. Fish and Wildlife Service and its National Wildlife Refuge System have always relied on sound science to inform management actions. Clearly, there are staffing and funding challenges, but scientific knowledge is the foundation of all conservation work. Conservation science in the National Wildlife Refuge System is guided by four foundational elements (USFWS, 2011):

- Apply science to refuge management to understand root causes, formulate hypotheses, reduce uncertainty, improve efficiency, and to solve complex problems;
- Complete robust inventory and monitoring to aid in the delivery of effective conservation;
- Develop deliberate research agendas that directly influence management decisions; and
- Expand communication, collaboration, and contribution within the U.S. Fish and Wildlife Service and among its partners to better understand and solve complex problems.

At the DRIWR, science has been a priority since its establishment in 2001. Selected examples of this commitment to science are presented in Table 15. It should be noted that this is not a comprehensive list, but only selected examples of scientific research, monitoring, and surveys being performed at DRIWR to properly inform management actions. Citizen science is also critically important to urban refuges and urban conservation initiatives. Every effort must be made to ensure that all monitoring and survey protocols used in citizen science are scientifically-accepted and approved by the refuge biologist, and that quality assurance/quality control practices are applied to data collected.

Table 15. Selected examples of scientific research, monitoring, and surveys performed at DRIWR to inform management actions.

Scientific Topic	**Example Publication**
Dragonfly and Damselfly surveys at Humbug Marsh	Craves (2007)
Christmas Bird Count monitoring	Craves (2006)
Humbug Marsh amphibian and reptile survey	Mifsud (2005)
Detroit River Hawk Watch	Stein et al. (2013)
Herring Gull (*Larus argentatus*) status and trends	Norwood (2011)
Common Tern (*Sterna hirundo)* assessment and management	Norwood et al. (2011)
Migratory bird stopover habitat in western Lake Erie	Ewert et al. (The Nature Conservancy, Lansing, MI, 2005)
Regional and landscape level movements of landbirds during migratory stopover in western Lake Erie	Dossman et al. (2013)
Cooperative weed management	Norwood and Stevens (2012)
Use of remote sensing and geographical information systems to better manage the Huron-Erie Corridor	Francoeur et al. (2012)
Mapping the invasive *Phragmites australis* on units of the Detroit River International Wildlife Refuge	Alessi et al. (2011)

Table 15. Cont'd

Detroit River-Western Lake Erie Indicator Project	Hartig et al. (2009)
Lake Whitefish (*Coregonus clupeaformis*) recovery	Roseman et al. (2012)
Walleye (*Sander vitreus*) and White Sucker (*Catostomus commersoni*) spawning in the Detroit River	Manny et al. (2010)
Lake Sturgeon (*Acipenser fulvescens*) population dynamics	Mohr et al. (2013)
Lake Sturgeon spawning in the Detroit River	Caswell et al. (2004)
Lake Sturgeon recovery	Roseman et al. (2011)
Lake Erie habitat	Weimer et al. (2012)
Soft shoreline engineering survey	Hartig et al. (2011)
Brownfield cleanup and habitat restoration at the Refuge Gateway	Hartig et al. (2012)

Decisions within the DRIWR are made within an adaptive management framework, where assessment is undertaken, management priorities are established, and management actions are taken in an iterative fashion for continuous improvement. Monitoring, research, and surveys are essential to practice adaptive management and without them management would be flying blind. Every effort should be made to ensure a strong coupling of this science to conservation management actions.

In urban refuges and urban conservation initiatives, like politics, you must also be prepared to compromise to make progress. It is useful to think about it in an adaptive management context. Urban management actions taken within an adaptive management context are often not comprehensive solutions. Frequently, additional actions are needed to fully resolve the problem and achieve the goal. Short-term actions, based on compromise, must be viewed in this continuous improvement model of adaptive management. A high priority must be placed on performing periodic rigorous quantitative assessments that can be used to measure continuous improvement through incremental steps. We must remember that the environmental and natural resource problems faced in major urban and industrial areas did not occur in a short period of time and will undoubtedly take considerable time to solve. Key questions to focus on to ensure science-based continuous improvement include:

- Has there been a proper scientific assessment?
- Is there a scientific rationale for the proposed conservation action?
- Are the conservation actions being taken going in the right direction (is the trajectory correct – toward long-term measurable goals)?
- Will there be follow-up ecological assessment and monitoring?
- Is there a commitment to adaptive management and therefore a process in place to make additional improvements and mid-course corrections, as necessary?

Recommendation: It is recommended that urban refuges and other urban conservation initiatives place a priority on the use sound science to properly inform management decisions and practice adaptive management.

LESSON 3: Develop Partnerships at All Levels and Don't Be Afraid to Experiment with Partnerships

In the urban conservation field, partnerships are critical because of the number of people and stakeholders involved and impacted, and the complexity of problems and solutions. Most urban areas, by their very nature, usually have high human population densities, substantial housing and commercial developments, urban sprawl, and at least some industrial development. Such urbanization and industrialization have frequently left a legacy of lost wildlife, wooded areas, and wetlands, and often a legacy of contaminated environments upon which our cities and their ecosystem health depend. To address these challenges and indeed bring conservation to urban areas will require both informal and formal public-private partnerships that are established through memoranda of understanding, partnership agreements, cooperative management agreements, seats on boards of directors and commissions, or other institutional mechanisms. Indeed, the conservation challenges in urban areas are far too big for any single agency, organization, or interest group to surmount – they require all to do their part and to work together.

These partnerships will have to undertake cooperative planning, share resources, seek external funds and often provide matching funds or in-kind services, cooperatively manage project delivery, and showcase the conservation work through creative outreach. Federal, state, and provincial agencies typically add value to these projects. Other key factors in successful urban conservation partnerships include: effective working relationships, trust among partners, clarity of roles and responsibilities, well recognized benefits to all partners, and effective facilitation.

Based on the experiences at DRIWR, urban partnerships should be project driven. Initial efforts should focus on recruiting and inspiring partners to get involved in a compelling conservation project that has benefits to all. Consider starting with simple projects like building a native garden, planting trees, removing invasive species, or other easily achievable one. Each project should be founded on cooperative learning and action. A core project delivery team is useful to coordinate among partners and to ensure that the project keeps on schedule, and to ensure that all partners fulfill commitments to the project. If the urban conservation project requires any permits, early efforts should be made to ensure that all regulatory agencies are involved up front in the process to ensure buy-in, support for project, and timely review and approval of permits.

The potential of urban partnerships to deliver conservation projects, key conservation messages, and rewarding stewardship experiences is enormous. Leaders can't be afraid to experiment with new partnerships. Indeed, one primary imperative should be to bring new and unconventional partners into the fabric of conservation in urban areas. This should be seen as institutional capacity-building for urban conservation. Experience in the greater Toronto metropolitan area has shown that successful conservation initiatives (Crombie, 1992):

- adopt a set of guiding principles (e.g. clean, green, connected, open, accessible, useable, diverse, affordable, and attractive);
- adjust plans to ensure they reflect these principles and an ecosystem approach;
- secure intergovernmental commitments, agreements, and commitments on what needs to be done, the priorities, partner responsibilities, and timeframe for action;
- consolidate capital budgets and pool resources, as necessary, to move projects forward;
- create the framework and conditions for private sector involvement, capitalizing on its enterprise, initiative, creativity, and capability for investment; and
- establish public-private partnerships.

There is another benefit of urban conservation partnerships. Typically, the amount of conservation land owned by any federal, state, provincial, or local agency, or any land trust or other nonprofit organization in a major urban area, is small. Therefore, much can be gained from working "beyond the boundaries" of refuges, state parks, metroparks, and other conservation areas to conserve entire landscapes. This will require working with nontraditional partners on private lands. Indeed, Chapter 5 provides excellent examples of how innovative urban partnerships have been used to work beyond refuge boundaries on over 50 soft shoreline engineering projects. This benefits fish and wildlife, and their requisite habitats, on landscape and ecosystem scales.

Recommendation: It is recommended that urban refuges and other urban conservation initiatives establish both informal and formal public-private partnerships for conservation projects founded on cooperative learning.

LESSON 4: Place a Priority on Developing a Land/Ecosystem Ethic Through Broad-Based Education, Outreach, and Stewardship

Urban residents must see themselves as part of the land/ecosystem. Both a land ethic and ecosystem approach call for people to recognize that they are part of the land/ecosystem and not separate from it. We need to reconnect people to the land and their ecosystem through cooperative learning (including service learning), playing, recreating, working, socializing, enjoying, and celebrating in their ecosystem.

Reconnecting to the land/ecosystem through a compelling outdoor experience can lead to thinking fresh about our relationship to our land/ecosystem. This can then lead to development of an ecosystem/land ethic that can inspire people to live differently. Living inspired by a land/ecosystem ethic gives hope. One of the goals should be to make sure that a land/ecosystem ethic becomes part of the urban community fabric. As Leopold (1949) noted, we must learn to love and respect the land, our ecosystem, and the place we call home. To love something requires passion. Loving something will also lead to respecting it and to becoming a good steward of it.

Education and service learning will be critical to developing a land/ecosystem ethic. No one person, organization, institution, or agency has all the answers. Answers and solutions will arise out of a cooperative learning process that involves urban stakeholders learning and working together to accomplish the common conservation goal, under conditions that involve positive interdependence and individual and group accountability. Such cooperative learning is essential to educate and inspire people to:

- understand problems, causes, and ramifications;
- address carrying capacity;
- conserve and enhance natural resources;
- protect the environment;
- foster a land/ecosystem ethic;
- ensure environmentally-sustainable economic development; and
- avoid the next "tipping point."

Everyone will be learning their way out. Urban refuges and other urban conservation initiatives must all become living laboratories for environmental education programs targeted at all age groups.

Recommendation: It is recommended that urban refuges and other urban conservation initiatives place a priority on providing compelling outdoor experiences that can lead to thinking fresh about urbanites' relationships to their land/ecosystem. This can then lead to development of a land /ecosystem ethic that can inspire people to live differently. Living inspired by a land/ecosystem ethic gives hope. One of the goals should be to make sure that a land/ecosystem ethic becomes part of the urban community fabric.

LESSON 5: Connect Urban Children and Families With Nature

Previous generations of kids played in the woods, along rivers, and in wetlands. They were building tree houses, catching fish, hunting frogs, and playing hide-and-seek. All of this "nature play" builds a sense of wonder and exploration, and enriches growth and development.

Over the past 40 years, more and more, children have been losing their connection to nature. Educators in childhood development now recognize that children require nature. There is a growing body of scientific evidence that suggests that if children are given early and ongoing positive exposure to nature, they thrive in intellectual, psychological, spiritual, and physical ways that their "shut in" peers do not.

By reducing stress, sharpening concentration, and stimulating creative problem-solving, "nature play" is also emerging as a promising therapy for attention-deficit disorder and other childhood maladies. Indeed, Richard Louv has argued compellingly in *Last Child in the Woods* that society needs a major effort to save our children from nature deficit disorder (Louv, 2005).

According to Louv (2008), this children and nature movement is being fueled by the fundamental idea that "the child in nature is an endangered species, and the health of children and the health of the Earth are inseparable." The trigger for this awakening has been the convergence of a number of trends, including:

- intensified awareness of the relationship between human well-being, the ability to learn, and environmental health;
- concern about child obesity; and
- media attention to nature-deficit disorder (Louv, 2008).

As a result of this awakening, there is much exciting work underway to reconnect children and families to the natural world.

From a parental and family perspective, the reasons we should take children outside are simple:

- It makes kids happier;
- It makes kids healthier;
- It makes kids smarter;
- It's free; and
- It's fun for the entire family.

A broad array of outdoor activities is needed to link a diverse urban citizenry with their surrounding natural world. Indeed, the mosaic of people living in urban areas will no doubt require a mosaic of tactics to connect urban children and families with nature.

We need unique urban conservation places, whether they be urban refuges, urban conservation areas, urban state parks, metroparks, city parks, conservancy lands, or other natural areas, or some combination of these urban conservation places, that can make nature experiences part of everyday urban life. Clearly, such urban conservation places have the potential to play an important role in addressing nature deficit through close-to-home, outdoor, recreational activities, hands-on outdoor learning opportunities, natural resource stewardship activities, nature-based health and fitness activities, and other innovative ideas. And these unique proximal natural resources can help children and families experience nature as the supporting fabric of their everyday lives. Whether it's hiking, fishing, hunting, birding, learning through environmental education, photography, environmental interpretation, or just plain exploring in the woods, urban refuges and other urban conservation areas have what educators, city planners, business leaders, and parents want – unique natural resources that can enhance quality of life, contribute to ecosystem health and healthful living, and nourish our sense of wonder, imagination, and curiosity.

In any major urban area, it is important that government agencies, educational institutions, businesses, environmental organizations, conservation clubs, faith-based organizations, and concerned citizens join forces in a coalition to help reconnect children and families to the land and water through compelling outdoor recreational and educational experiences (i.e. learning and service learning, playing, exploring, and recreating) that help foster an appreciation and love for the outdoors. That, in turn, will help develop a strong sense of place that inspires positive actions, a sense of ownership, and stewardship for the community's natural resources. In any urban area, it would be beneficial to develop and nurture an active and vibrant friends' group or community partnership to help connect urban children and families with nature.

Recommendation: It is recommended that unique urban conservation places, like city parks, metroparks, urban state parks or refuges (if present), greenway trail systems, or other urban conservation areas, or a network of such areas, offer: compelling outdoor recreational activities; hands-on outdoor learning opportunities; natural resource stewardship activities; nature-based health and fitness activities; and other innovative ways and means to make nature experiences part of everyday urban life.

LESSON 6: Build a Record of Conservation Success and Celebrate It Frequently

In any major urban conservation initiative, particularly when the population affected and number of stakeholder groups are large, it is important to build a record of conservation success and celebrate it frequently in a very public fashion. In the spirit of adaptive management, natural resource managers assess status, set priorities, and take conservation actions in an iterative fashion for continuous improvement. Each assessment, whether performed annually or biennially, should compile all conservation successes and communicate their benefits to urban residents. This could be an annual or biennial registry of conservation accomplishments that informs, educates, and inspires more urban residents to get involved, and to help develop the next generation of conservationists. It is useful to make sure that the conservation accomplishments and benefits get recorded in a fashion that is meaningful to the public and resonates with them. People always like to hear how a conservation or community action affects them directly. Use the media to help get these messages out.

Urban refuges and urban conservation initiatives should develop an effective communications' plan that tells their compelling conservation story to urban audiences, using all available communication technologies to get the message out. Further, priority should be placed on cultivating the media and involving politicians, prominent community and business leaders, and school systems in celebrating the conservation accomplishments. With the disconnection between urban residents and nature, the complexity of issues facing urban areas, and the competing priorities of urban residents, the conservation community must routinely be out in front of the urban citizenry with messages and stories of relentless positive conservation actions.

Recommendation: It is recommended that urban refuges and other urban conservation initiatives build a record of conservation success and celebrate it frequently in a very public fashion.

LESSON 7: Quantify Economic Benefits

By any measure, the value of our great outdoors is impressive. For example, the most recent economic study of economic benefits of national wildlife refuges (Carver and Caudill, 2013) showed what most hunters, anglers, birders, and conservationists already knew – that wildlife refuges are over a billion dollar business. This study, titled *Banking on Nature: The Economic Benefits to Local Communities of National Wildlife Refuge Visitation,* found that recreational use on national wildlife refuges generated almost $2.4 billion in sales and economic output during 2011 (Carver and Caudill, 2013). In total, 46.5 million people visited national wildlife refuges in 2011, supporting 35,000 private sector jobs and producing about $792.7 million in employment income. In addition, recreational spending on refuges generated nearly $342.9 million in tax revenue at the local, county, state and federal level. These economic benefits are nearly five times the amount appropriated to the National Wildlife Refuge System in Fiscal Year 2006. Carver and Caudill (2013) found that about 77% of refuge visitors in 2013 travelled from outside the local area.

In addition, a peer-reviewed study performed by North Carolina State University researchers for the U.S. Fish and Wildlife Service showed that a home located near a national wildlife refuge increases its value and helps support the surrounding community's tax base (Taylor et al., 2012). This study, based on 2000 U.S. Census Bureau data, found that the value of homes located within 0.8 km (one half mile) of a national wildlife refuge and within 12.9 km (8 miles) of an urban center were valued 5-9% higher than those that did not meet those criteria (Taylor et al., 2012).

These economic benefits data from wildlife refuges are supported by economic benefits studies of Great Lakes cleanup and restoration. For example, Austin et al., (2007) have estimated that a $26 billion investment in cleanup of the Great Lakes through the Great Lakes Regional Collaboration would result in $50 billion in long-term economic benefits.

In downtown Detroit, the Detroit Riverfront Conservancy completed an economic benefits study of industrial waterfront reclaimed as public greenspace for nearly three million annual visitors (i.e. the Detroit RiverWalk and associated green infrastructure). As of 2012, 80% of the east Detroit RiverWalk was constructed at a cost of $80 million, coupled with a $60 million endowment for operation and maintenance (CSL International, 2013). This investment was a catalyst for $1.55 billion in total public and private sector investment (including the value of contributed land), of which approximately $639 million can be directly linked to riverfront improvements (CSL

International, 2013). In addition, there is potential for an additional $700-950 million investment in the future (CSL International, 2013).

Using these three economic studies, the return on investment in wildlife refuges (5 to 1), Great Lakes cleanup (2 to 1), and Detroit RiverWalk (nearly 5 to 1) provide compelling evidence for additional investment in these programs. Further, such compelling economic rationale helps focus attention on the importance of conservation and natural resource restoration, and helps attract future conservation partners like businesses, foundations, and others, particularly in urban areas. Therefore, quantitative economic benefits assessments should be performed for urban refuges and urban conservation initiatives to help build a record of conservation success and celebrate it frequently in a very public fashion.

Recommendation: It is recommended that urban refuges and other urban conservation initiatives quantify economic benefits of conservation and outdoor recreation to help focus attention on their importance, and to help attract future conservation partners like businesses, foundations, and others.

LESSON 8: Involve the Public in All Conservation Actions to Develop a Sense of Place, Establish Local Ownership, and Instill Local Responsibility for Stewardship

Urban wildlife refuges and other urban conservation areas provide unique opportunities for people to become involved in a positive way with the natural world that can inspire them to take personal responsibility for helping with the stewardship of these natural resources. First, urban wildlife refuges and other urban conservation areas provide unique opportunities for outdoor recreation like fishing, hunting, wildlife observation, wildlife photography, environmental education, environmental interpretation, canoeing, kayaking, and more. Through compelling outdoor recreational and educational experiences citizens can gain a better appreciation and love for the outdoors. That, in turn, will help develop a strong sense of place that inspires positive actions, a sense of ownership, and stewardship for these urban natural resources.

Second, the large urban constituencies adjacent to urban refuges and urban conservation areas are generally receptive to participating in service learning and stewardship experiences that are located in their own community. From the very beginning, one of the goals of the DRIWR has been to create a unique urban refuge that provides an exceptional conservation experience and to do it in a fashion that grows hope and inspires the next generation of conservationists and sustainability entrepreneurs. The brownfield cleanup and habitat restoration work at the Refuge Gateway (home of the Refuge's visitor center under construction) provided a unique opportunity to involve all stakeholders, from a wide range of ages and backgrounds, in all phases of this process, including planning, design, fundraising, implementation, and stewardship. Again, it was hoped that a sense of ownership would develop and lead to a commitment of stewardship. Attempts to foster feelings of place-attachment and stewardship (which can develop from recognition of a sense of place) were made throughout all phases of the project through the following strategies:

- use of local materials to give built features a unique physical appearance that fits well into its environment;
- involvement of local volunteers and labor forces;
- site-specific design of public-use features; and
- place-based educational activities conducted throughout all phases of construction (Hartig et al., 2010).

By showcasing the unique attributes of the Refuge Gateway and adjacent Humbug Marsh through environmentally-sensitive and place-based design, and involvement of stakeholders in stewardship and restoration projects, a sense of place, local ownership, and stewardship of the land is slowly being cultivated. From tree and shrub plantings, and invasive species removal days, through business team-building projects and high school and university field trips, emphasis was consistently placed on citizen/student contributions to caring for their refuge, creating a feeling of local ownership and a sense of place. Indeed, many of the volunteers have come back to the refuge after their experience and brought their friends and families to show what they had done in a sense of personal accomplishment and pride. It should also be noted that establishing a sense of place and local ownership helps develop citizen leaders of conservation.

Recommendation: It is recommended that urban refuges and other urban conservation initiatives place a high priority on involving urban residents in all urban conservation activities to develop a sense of place and local ownership, and to instill local responsibility for stewardship.

LESSON 9: Recruit and Train Staff to be Urban Change Agents and Facilitators

It is not just anyone who can work in an urban wildlife refuge or on an urban conservation initiative. First, they need to clearly understand their important dual role of conserving fish and wildlife, and their habitats, and reaching and teaching the next generation of conservationists. They need to be able to relate to urban people and clearly understand the barriers and challenges to reaching them. Ideally, they need to be part of the local community so that they understand it and can work within it, or be willing to become part of it. Ideally, they need to have established community relationships and trust, or be willing to establish them.

People working on urban conservation need to be good facilitators of partnerships across the full spectrum of diversity (e.g. culture, age, economic status, religion, etc.) and they should have experience in establishing and sustaining public-private-nonprofit partnerships (USFWS, 2011). They should also have the skills to organize and manage an institutional framework that is unique to the area (USFWS, 2011).

In essence, the urban conservation staff needs to be urban conservation change agents. The kind of staff needed for wilderness, rural, and agricultural conservation programs isn't necessarily the best for delivering conservation in urban areas.

A change agent is a person that acts as a catalyst and champion for change, and then plays a key role in planning and managing its implementation. They are individuals whose presence and/or thought processes cause a change from a traditional way of thinking about or handling a management issue or problem. It has been said that a change agent:

- lives in the future, not the present – they don't see something like it is, they see it for what it can be;
- is passionate and inspires passion in others;
- understands people and the context or place within which they live and work; and
- is self-motivated and a good leader of people and multi-stakeholder processes.

Urban conservation change agents must have the professional acumen, the people skills, and the ability to earn the respect of urban residents and all stakeholder groups. These change agents need to be exceptional communicators, adaptable, innovative, transparent, calculated risk-takers, partnership-builders, and clearly passionate about developing the next generation of conservationists. Undoubtedly, these urban conservation change agents need to be recruited and trained specifically for becoming leaders and core delivery team members of urban refuges and other urban conservation initiatives.

Again, urban refuge and urban conservation initiative change agents must be passionate about both protecting and conserving fish and wildlife, and their habitats, and about bringing nature and conservation into the urban area in an effort to develop the next generation of conservationists. It should also go without saying, that adequate resources will be essential to deliver urban conservation effectively. It has been said that budget manifests mission. If it is part of the conservation mission to bring natural resource experiences to urban areas because that is where most people live, than adequate human and financial resources must be provided.

Recommendation: It is recommended that urban conservation leaders and core delivery team members be recruited and trained to become change agents that creatively conserve fish and wildlife, and their habitats, and creatively make nature experiences part of everyday urban life.

LESSON 10: Recruit a High-Profile Champion

From the very outset, the DRIWR was very fortunate to have several high-profile champions:

- former corporate executive of Stroh Companies, Incorporated and Greater Detroit American Heritage River Initiative Chairman Peter Stroh;
- then Canadian Deputy Prime Minister Herb Gray, and
- the Dean of the United States Congress – John D. Dingell.

Deputy Prime Minister Herb Gray lived, at the time, in Windsor, Ontario along the Detroit River. Peter Stroh was a third generation executive of Stroh Brewing Company, a very well-respected businessperson, and an avid outdoorsperson and conservationist who served as a trustee for numerous international conservation organizations. Congressman John Dingell was the longest-serving member of the U.S. House of Representatives, an avid outdoorsperson and conservationist, and an author of the Clean Water Act, the Endangered Species Act, the Marine Mammal Protection Act, and many other important laws. Collectively, these three men lent their names and reputations to the urban refuge effort and gave it legitimacy.

Each of these men was well-respected in the region, cared deeply about it, recognized the importance of the Detroit River and western Lake Erie to the region's economy, communities, and quality of life, and wanted to do something special for the region. Each of these men, because of his stature, could help get media attention and open doors.

Peter Stroh used his influence as a well-respected businessperson to open corporate doors. Unfortunately, Peter Stroh passed away in 2002 and Herb Gray went on to serve as Canadian Chairman of the International Joint Commission before being appointed Chancellor of Carleton University in Ottawa, Ontario in 2008. DRIWR is now fortunate to have current Canadian Member of Parliament Jeff Watson serving as a high profile champion on the Canadian side. Congressman John Dingell continued to serve as the refuge's high profile champion on the U.S. side until his retirement in December 2014. He has:

- brought people together in the spirit of cooperative conservation;
- opened doors;
- helped overcome institutional obstacles; and
- built the capacity for a world-class international wildlife refuge in a major urban and industrial area.

Clearly, there is no doubt that the DRIWR would not be where it is today or have accomplished what it has without the passion and leadership of Congressman John Dingell. Although it is probably not essential to have high profile champion for an urban refuge or urban conservation initiative, it clearly helps enormously and pays huge dividends.

Recommendation: It is recommended that urban refuges and other urban conservation initiatives recruit a well-recognized conservation champion to help raise public awareness, bring people and organizations together in the spirit of cooperative conservation, open doors, overcome institutional obstacles, and build capacity for urban conservation.

Concluding Thoughts

The percentage of people in the world living in urban areas has increased from 29% in 1950 to 49% in 2007 and is projected to increase to 60% by 2030 (United Nations 2006; Haub, 2007). Today, approximately 80% of all people in the U.S. and Canada live in urban areas. Most urban residents are disconnected from the natural world. As a global community, we cannot afford to allow this disconnection to continue. In addition, urban areas are places where most biologists, conservation agencies, and conservationists have historically shied away from. This compounds the problem of bringing conservation to urban areas and of helping develop the next generation of conservationists in urban areas because that is where most of the people on our planet live.

There is no doubt that urban refuges and other urban conservation initiatives have the potential to make a significant impact on developing the next generation of conservationists. Urban areas also provide unique opportunities to form public-private partnerships and to leverage resources for conservation. Indeed, that is precisely what is happening at the DRIWR that has worked with over 300 public and private organizations, and leveraged over $43 million for conservation projects since 2001.

Today, we stand at a new urban conservation frontier, one that has numerous challenges and opportunities. We need to commit to exploring this frontier and charting a course that brings conservation into urban areas in a fashion that makes nature experiences part of everyday urban life. As a global community, we cannot afford to fail.

Literature Cited

Alessi, J., Jaworski, E., Burke, J., 2011. Mapping the Invasive *Phragmites australis* on Units of the Detroit River International Wildlife Refuge. State of the Strait Conference, 2011. Eastern Michigan University, Ypsilanti, Michigan, USA.

Austin, J.C., Anderson, S., Courant, P.N., Litan, R.E., 2007. Healthy Waters, Strong Economy: The Benefits of Restoring the Great Lakes Ecosystem. Brookings Institution, Washington, D.C., USA.

Carver, E., Caudill, J., 2013. Banking on Nature: The economic benefits to local communities of national wildlife refuge visitation. U.S. Fish and Wildlife Service, Division of Economics, Washington, D.C., USA.

Caswell, N.M., Peterson, D.L., Manny, B.A., Kennedy, G.W., 2004. Spawning by Lake Sturgeon (*Acipenser fulvescens*) in the Detroit River. Journal of Applied Ichthyology. 20, 1–6.

Craves, J. A., 2006. Thirty years of the Rockwood Christmas Bird Count, 1975-2004. Report to the U.S. Fish and Wildlife Service, Great Lakes Coastal Program. East Lansing, Michigan, USA.

Craves, J. A., 2007. Baseline inventory of Odonata at the Detroit River International Wildlife Refuge, Humbug Marsh Unit. Final Report, CCS MOA #2007 CCS-98. U.S. Fish and Wildlife Service, Region 3, Fort Snelling, Minnesota, USA.

Crombie, D., 1992. *Regeneration: Toronto's Waterfront and the Sustainable City.* Queen Printer of Ontario, Toronto, Ontario.

CSL International. 2013. Economic impact study: Detroit riverfront. Detroit Riverfront Conservancy, Detroit, Michigan, USA.

Dossman, B, Rodewald, P.G., Matthews, S., 2013. Regional and landscape level movements of landbirds during migratory stopover in western Lake Erie. The Ohio State University, School of Environment and Natural Resources, Columbus, Ohio, USA.

Francoeur, S., Cargnelli, L., Cook, A., Hartig, J., Gannon, J., Norwood, G. (Eds.), 2012. State of the Strait: Use of Remote Sensing and GIS to Better Manage the Huron-Erie Corridor. Great Lakes Institute for Environmental Research, Occasional Publication No. 7, University of Windsor, Ontario, Canada.

Hartig, J.H., Zarull, M.A., Ciborowski, J.J.H., Gannon, J.E., Wilke, E., Norwood, G., Vincent, A., 2009. Long-term ecosystem monitoring and assessment of the

Detroit River and western Lake Erie. Environmental Monitoring and Assessment. 158, 87-104.

Hartig J.H., Robinson, R.S., Zarull, M.A., 2010. Designing a Sustainable Future through Creation of North America's only International Wildlife Refuge. Sustainability. 2(9), 3110-3128.

Hartig, J.H., Zarull, M.A., Cook, A., 2011. Soft shoreline engineering survey of ecological effectiveness.Ecological Engineering. 37, 1231-1238.

Hartig J.H., Krueger, A., Rice, K., Niswander, S., Jenkins, B., Norwood, G., 2012. Transformation of an industrial brownfield into an ecological buffer for Michigan's only Ramsar Wetland of International Importance. Sustainability 4(5), 1043-1058.

Haub, C., 2007. *World population data sheet*. Population Reference Bureau, Washington, D.C., USA.

Kareiva, P., 2013. Urban conservation: Conservation should be a walk in the park, not just a walk in the woods. The Nature Conservancy.

Leopold, A., 1949. *A Sand County Almanac: And Sketches Here and There*. Oxford University Press, New York, New York, USA.

Louv, R., 2005. *Last Child in the Woods: Saving Our Children from Nature-Deficit Disorder*. Algonquin Books of Chapel Hill, Chapel Hill, North Carolina, USA.

Louv, R., 2008. Child and nature movement. Richard Louv Website. http://richardlouv.com/books/last-child/children-nature-movement/ (April 2013).

Manny, B.A., Kennedy, G.W., Boase, J.C., Allen, J.D., Roseman, E.F., 2010. Spawning by Walleye (*Sander vitreus*) and white sucker (*Catostomus commersoni*) in the Detroit River: Implications for spawning habitat enhancement. Journal of Great Lakes Research. 36:490-496.

Metropolitan Affairs Coalition.2001, A Conservation Vision for the Lower Detroit River Ecosystem. Detroit, Michigan, USA (http://www.fws.gov/uploadedFiles/a%20conservation%20vision%20for%20the%20lower%20detroit%20river%20ecosystem.pdf (May2014).

Mifsud, D., 2005. Humbug Marsh Amphibian and Reptile Survey. Herpetological Resource and Management, Grass Lake, Michigan, USA.

Norwood, G., Schneider, T., Jozwiak, J., Cook, A., Hartig, J.H. (Eds.), 2011. Establishing a quantitative target for common tern management at Detroit River and western Lake Erie. Final report to U.S. Environmental Protection Agency. Chicago, Illinois, USA.

Norwood, G., 2011. Herring Gull *(Larus argentatus)*. In: A.T. Chartier, J.J. Baldy, J.M. Brenneman (Eds.), *The Second Michigan Breeding Bird Atlas*. Kalamazoo Nature Center. Kalamazoo, Michigan, USA.

Norwood, G., Stevens, G., 2012. Detroit River-Western Lake Erie Cooperative Weed Management Area Annual Report. Detroit River International Wildlife Refuge, Grosse Ile, Michigan, USA.

Roseman, E.F., Manny, B., Boase, J., Child, M., Kennedy, G., Craig, J., Soper, K., Drouin, R., 2011. Lake Sturgeon response to a spawning reef constructed in the Detroit River. J. Appl. Ichthyol. 27 (Suppl. 2), 66–76.

Roseman, E.F., Kennedy, G., Manny, B.A., Boase, J., McFee, J., 2012. Life history characteristics of a recovering Lake Whitefish *Coregonus clupeaformis* stock in the Detroit River, North America. Proceedings of the 10th Coregonid Fishes Symposium. Advances in Limnology 63, 477-501.

Rybczynski, W., 1999. *A Clearing in the Distance: Frederick Law Olmstead and America in the 19th Century*. Touchstone, New York, New York, USA.
Senge, P.M., 1990. *The Fifth Discipline: The Art and Practice of the Learning Organization*. Currency Doubleday Books, New York, New York, USA.
Stein, J., Norwood, G., Conard, R., 2013. Detroit River Hawk Watch 2012 Season Summary. U.S. Fish and Wildlife Service, Detroit River International Wildlife Refuge, Grosse Ile, Michigan, USA.
Taylor, L.O., Liu, X, Hamilton, T., 2012. Amenity Values of Proximity to National Wildlife Refuges. U.S. Fish and Wildlife Service, Washington, D.C., USA.
United Nations, 2006. World urbanization prospects: The 2005 revision. New York, New York, USA.
USFWS, 2011.*Conserving the Future: Wildlife Refuges and the Next Generation*. National Wildlife Refuge System, Arlington, Virginia, USA.
Weimer, E., MacDougall, T., Gorman, A.M., Boase, J., Francis, J., Geddes, C., Haas, B., Kenyon, R., Knight, C., Kocovsky, P., Roseman, E., Mackey, S., Markham, J., Rutherford.E., 2012. Lake Erie Habitat Task Group Report to the Lake Erie Committee, Great Lakes Fishery Commission, Ann Arbor, Michigan, USA.

Epilogue

> *Wild beasts and birds are by right not the property merely of the people who are alive today, but the property of unknown generations, whose belongings we have no right to squander.*
>
> **President Theodore Roosevelt**

A surprise is a feeling of sudden wonder or amazement at something unexpected. It is my hope that you have been pleasantly surprised to learn that the Detroit River is no longer one of the most polluted rivers in North America, that its environmental improvements have led to one of the most dramatic ecological recoveries in North America, and that now, along with western Lake Erie, it is part of the only international wildlife refuge in North America and an experiment in bringing nature and wildlife into an urban industrial area where millions of people live. This recovery is remarkable, but much remains to be done to meet the long-term goals of restoring the physical, chemical, and biological integrity of this ecosystem, and to meet the long-term goals of the Detroit River International Wildlife Refuge (DRIWR), including developing the next generation of conservationists.

I like to think of the DRIWR as a tapestry. A tapestry is a form of textile art, traditionally woven on a vertical loom and most often proudly displayed in a prominent location of a home. Individual colored threads, each unique and beautiful in their own right, are woven together to produce an exceptional piece of art more beautiful and much stronger than imagined with just the individual threads. The DRIWR is like an ecological tapestry made up of numerous species and habitats that when woven together are more beautiful and much stronger than imagined with just the individual species and habitats. Much like a textile tapestry is a source of pride in the home, the DRIWR tapestry has

> *We did not weave the web of life; we are merely a strand in it. Whatever we do to the web, we do to ourselves.*
>
> **Chief Seattle**

become a source of pride in southeast Michigan and southwest Ontario.

There is no scientific doubt that the DRIWR would truly be unique in its own right, because of its plethora and diversity of fish and wildlife, if it were not situated in the industrial heartland and a nearly seven-million person urban area. But it is and, just like a rose that grows surrounded by concrete and steel is more remarkable than one that grows in a horticulturist's garden, the DRIWR is more remarkable because it is being built in the industrial heartland and within a major urban area.

> *When one tugs at a single thing in nature he finds it attached to the rest of the world.*
>
> **John Muir**

From 2.1 m (seven foot) Lake Sturgeon to 14 g (0.5 ounce) warblers, the DRIWR is a priceless gift to us and future generations. Indeed, the future of the DRIWR is very promising. If the lessons in Chapter 10 are followed, it is realistic to see the DRIWR grow to over 10,100 ha (25,000 acres) by 2021 – the 20-year anniversary of refuge establishment. Can you imagine 10,100 ha of land and water devoted to conservation in the automobile capitals of Canada and the United States, and in a nearly seven million person urban area? These lands and waters, and their continentally-significant biodiversity, have intrinsic value and provide ecosystem services that enhance competitive advantage for communities and businesses, and that increase quality of life and community pride. This urban industrial refuge is a unique anchor for biodiversity in the Great Lakes Basin Ecosystem and a model for urban and transboundary conservation initiatives throughout the world.

One of the critical factors necessary for realizing this predicted growth in conservation lands and reaping its associated benefits and ecosystem services is to continue to grow public-private partnerships for conservation, especially recruiting, equipping, motivating, and inspiring new partners. In life, most successful activities require effective relation-

> *When we see land as a community to which we belong, we may begin to use it with love and respect.*
>
> **Aldo Leopold**

ships between and among partners. Public-private partnerships for conservation are no different. Conservation organizations and agencies must establish meaningful relationships with all stakeholder groups that run deep and become part of the community fabric. Some of these public-private partnerships will be informal, while others will be formalized through memoranda of understanding, partnership agreements, cooperative management agreements, seats on boards of directors and commissions, or other institutional mechanisms. The key is to develop, nurture, and sustain strong and effective partnerships.

> *Every one of us makes a difference every day. Everyone has a role to play.*
>
> **Dr. Jane Goodall**

One useful way to think about public-private partnerships is through the ripple effect. When you throw a pebble into a pond, you can watch the ripples spread out across the water. It is always amazing that one small pebble can create such a wide circle. Just like a pebble creates a wide circle when thrown into a pond, we need a new generation of urban conservationists to create wide circles of conservation influence and impact through innovative partnerships. We need urban change agents that show us the way to becoming a conserver society and a sustainable community.

There will always be people who will say that urban environmental and natural resource problems are just too big to make a real difference. There is no doubt that the scope of urban problems, like urban sprawl, habitat loss and fragmentation, nonpoint source pollution, legacy pollution of contaminated sediments and industrial brownfields, and others, is substantial. However, we must have a sense of urgency to "get on with the job" before we reach another tipping point. Most people can relate to this through Lauren Eisley's story called "The Star Thrower" (Eisley 1969). In this story a man is walking down the beach when he notices a young boy

> *It is a wholesome and necessary thing for us to turn again to the earth and in the contemplation of her beauties to know of wonder and humility.*
>
> **Rachel Carson**

picking up starfish and throwing them into the ocean. The man approaches the boy and asks "What are you doing?" The boy replies: "Throwing starfish back into the ocean. The surf is up and the tide is going out. If I don't throw them back, they'll die." "Son," the man said, "don't you realize there are miles and miles of beach and hundreds of starfish? You can't make a difference!" After listening politely, the boy bent down, picked up another starfish, and threw it back into the surf. Then, he looked up, smiled at the man, and said "I made a difference for that one." It is precisely this philosophy that is needed in urban conservation work, where small incremental actions or steps are needed to complete a long journey.

It is critically important that a high priority be placed on reconnecting urban residents with nature as part of a long-term strategy to inspire individual respect, love, and stewardship of the land/ecosystem to be able to develop a societal land/ecosystem ethic for sustainability. All stakeholder groups, including government agencies, educational institutions, businesses, environmental organizations, conservation clubs, faith-based organizations, social advocacy groups, and health institutions, must join forces to help reconnect people to the land and water in urban areas through compelling outdoor recreational and educational experiences that help foster an appreciation of and love for the outdoors. That, in turn, will help develop a strong sense of place that inspires positive actions, a sense of ownership, and stewardship for the community's natural resources.

Volunteers will always be critically important to wildlife refuges and conservation initiatives, but the potential in urban areas is enormous. These volunteers need to both be inspired to get involved in citizen science and stewardship, and to become the next ambassadors of conservation in their urban area. Collectively, these urban volunteers need to be self-sustaining.

It is abundantly clear that urban refuges and other urban conservation places have the unique proximal natural resources to help children

> *Everyone needs beauty as well as bread, places to play in and pray in, where nature may heal and cheer and give strength to the body and soul.*
>
> **John Muir**

experience nature as the supporting fabric of their everyday lives. Whether it's hiking, fishing, hunting, birding, learning through environmental education, photography, natural resource interpretation, or just plain exploring in the woods, urban refuges and urban conservation areas have what educators, city planners, developers, business leaders, and parents want – unique natural resources that can enhance quality of life, contribute to ecosystem health and healthful living, and nourish our sense of wonder, imagination, and curiosity. And in the case of the DRIWR, these natural resources can be seen, enjoyed, and studied in the shadows of industries and skyscrapers, providing a foretaste of sustainable development that strives for balanced and continuous social, economic, and environmental progress. Urban wildlife refuges and other urban conservation initiatives must all become living laboratories for interpretive and environmental education programs targeted at all age groups.

We need unique urban conservation places, whether they be urban refuges, urban conservation areas, urban state parks, metroparks, city parks, conservancy lands, or other natural areas, or some combination of these urban conservation places, that can make nature experiences part of everyday urban life. It was indeed quite prophetic that the great American author and naturalist Henry David Thoreau, while watching civilization expand into the countryside during his lifetime (1817-1862), recommended that every town should have a forest of 202-404 ha (500-1,000 acres) to be used for conservation instruction and outdoor recreation (Malnor and Malnor, 2009). There is no doubt that unique urban conservation places will undoubtedly be part of every successful sustainable city in the future. Indeed, Cohen (1999) has shown that nature plays a key role in urban life, including preserving a city's structure and ensuring its well-being and success.

> *In merging nature and culture the most successful cities combine such universal needs as maintaining or restoring contact with the cycles of nature, with specific, local characteristics.*
>
> **Sally A. Kitt Chappell**

That is why it is now very exciting to see how the U.S. Fish and Wildlife Service (2011) has recognized, through its new vision document titled *Conserving the Future: Wildlife Refuges and the Next*

Generation, the urgent need to connect an ever growing and disconnected urban population to natural resources and, as a result, has established an "Urban Refuge Initiative" to address this need. In the future, more emphasis will be placed on partnerships and collaboration than on traditional refuge establishments. This Initiative has undertaken critical analysis of conservation relevance in urban areas and of creating a connected urban conservation constituency. Out of this critical analysis, the U.S. Fish and Wildlife Service (2013) identified that excellence in urban wildlife refuges is achieved when the following standards are met:

- Connect urban people with nature using a full spectrum of opportunities for "nature novices" through in-depth adventurous programs;
- Build partnerships;
- Become a community asset;
- Ensure adequate long-term resources;
- Provide equitable access;
- Ensure that visitors feel safe and welcome; and
- Showcase sustainability in all actions and activities.

Through this Initiative, the U.S. Fish and Wildlife Service established eight urban wildlife refuge partnerships across the United States in 2013 (i.e. New Haven Urban Wildlife Refuge Partnership, Connecticut; Forest Preserves of Cook County Urban Wildlife Refuge Partnership, Illinois; Albuquerque Urban Wildlife Refuge Partnership, New Mexico; Houston Urban Wildlife Refuge Partnership, Texas; Providence Parks Urban Wildlife Refuge Partnership, Rhode Island; Masonville Cove Urban Wildlife Refuge, Maryland; Lake Sammamish Urban Wildlife Refuge Partnership, Washington; and L.A. River Rover Urban Wildlife Refuge Partnership, California). Conservation in urban areas is not only a question about the amount and uniqueness of natural resources, it is a question about making nature experiences part of everyday urban life to help develop a conservation ethic. Really, it is a question of human heart.

The DRIWR is a unique urban place where the tapestry of life has been woven with elegance, where the music of life has been rehearsed to perfection for thousands of years, where nature's colors are most vibrant and engaging, where time is measured in seasons, and where the courtship dance of diving ducks takes center stage (USFWS 1999). It is a gift given to us for our appreciation, enjoyment, and inspiration, but also with a responsibility for stewardship so that it can be passed on to future generations. It is a gift unwrapped each time a hunter sets the decoys, an angler lands a fish, an amateur photographer clicks the shutter, a birder lifts binoculars, a paddler launches a kayak, and a child catches a tadpole (USFWS 1999).

> *I am here to speak for the generation to come.....I'm only a child, yet I know we are part of a family, five billion strong, in fact, 30 million species strong; and borders and governments will not change that. I am only a child, yet I know we are all in this together and should act as one single world toward one single goal.*
>
> **12-year old Severn Suzuki at the Earth Summit in Rio de Janeiro in 1992**

If conservation can be brought into the industrial heartland, the automobile capitals of the United States and Canada, and a major urban area with nearly seven million people through the work of the DRIWR, it can be done elsewhere. Urban conservation work is not easy and not for the faint of heart. It is frequently underappreciated. However, it is so important, much needed, and very rewarding. It is my hope that, in some small way, this book helps inspire the next generation of conservationists that must be developed with increasing frequency in major urban areas because that is where most people on our planet live.

Literature Cited

Cohen, N., 1999. *Urban Conservation*. The MIT Press, Cambridge, Massachusetts, USA.

Eisley, L., 1969. *The Unexpected Universe*. Harcourt Brace & Company, Orlando, Florida, USA.

Malnor, B., Malnor, C.L., 2009. *Champions of the Wilderness*. Dawn Publications, Nevada City, California, USA.
U.S. Fish and Wildlife Service (USFWS), 1999. Fulfilling the Promise: Visions for Wildlife, Habitat, People, and Leadership. National Wildlife Refuge System, Arlington, Virginia, USA.
USFWS, 2011. Conserving the Future: Wildlife Refuges and the Next Generation. National Wildlife Refuge System, Arlington, Virginia, USA.
USFWS, 2013. Introduction to the standards of excellence for urban national wildlife refuges. Washington, D.C., USA.

Glossary

Adaptive management = a management process that assesses, set priorities, and take action in an iterative fashion for continuous improvement.

Algae = simple one-celled or many-celled plants without true tissues, can be free-floating (e.g. phytoplanktonic), and capable of carrying on photosynthesis in aquatic ecosystems.

Algal bloom = a high density of algae, typically blue-green algae, that are often seen as free floating mats; they can occur suddenly as a result of opportunistic growth spurts and can adversely affect water quality.

Angler = a person who fishes with a rod and line.

Area of Concern (AOC) = a specific geographic area of the Great Lakes that fails to meet the objectives of the Great Lakes Water Quality Agreement where such failure has caused or is likely to cause one or more of 14 beneficial uses.

Armor stone = a loose assemblage of broken stones erected in water or on soft ground as a foundation; rock of varying sizes used to armor shorelines, streambeds, bridge abutments, pilings, and other shoreline structures against scour, water, or ice erosion.

Biodiversity = the number and variety of plant and animal species that exist in a particular ecosystem.

Biomagnification = a cumulative increase in the concentration of persistent toxic substances in successively higher trophic levels of the food web (i.e. from algae to zooplankton to fish to birds to humans).

Biota = the sum total of all living things (plants, animals, fungi) that inhabit a region or area.

Biological diversity = the total number of species of plants and animals in a given area or ecosystem.

Bio-stabilization = the use of erosion control materials such as erosions mats, filter fabrics, and/or rocks, combined with native plantings for the purpose of stabilizing slopes, stream banks, and river and lake shorelines.

Breakwater = a concrete wall or other structure designed to protect coastal land and waterfront developments from the battering of the waves and long-shore currents.

Brownfield = A tract of land that was originally developed for industrial purposes, contaminated, and then abandoned.

Cantilever = A projecting structure, such as a beam, that is supported at one end and carries a load at the other end or along its length.

Capping = in the context of brownfields, it is covering a contaminated area with clean material to minimize exposure to contaminants.

Change agent = a person from inside or outside the organization who helps an organization transform itself by focusing on such matters as organizational effectiveness, improvement, and development.

Charismatic megafauna = large animal species with widespread popular appeal that environmental activists use to achieve environmentalist goals.

Chicago Wilderness = an alliance of more than 260 public and private organizations that work together to understand, protect, restore, and better manage the natural ecosystems of the Greater Chicago region.

Chironomids = mosquito-like insects commonly called midges; their larvae live in the sediments of all types of aquatic habitats.

Cholera = an acute, diarrheal illness caused by infection of the intestine with the bacterium *Vibrio cholerae*.

Christmas Bird Count = an annual bird census conducted by birders around Christmas each year, representing the longest citizen science survey in the world.

Citizen science = scientific research and monitoring conducted, in whole or in part, by amateur or nonprofessional scientists.

Clean Water Act = the U.S. Federal Water Pollution Control Act, initially signed in 1972 and amended numerous times in an effort to establish national programs intended to address all types of water pollution control programs.

Combined sewer overflow = in sewerage systems that carry both sanitary sewage and storm water runoff, the portion of the flow which goes untreated to receiving streams or lakes because of wastewater treatment plant overloading during storms.

Comprehensive conservation plan = a document to guides long-term management of a wildlife refuge within the U.S. National Wildlife Refuge System.

Confined disposal facility = a structure for disposing of contaminated dredged spoils.

Consent Decree = an agreement or settlement to resolve a dispute between two parties without admission of guilt.

Conservation = the careful use and stewardship of natural resources to prevent them from being lost or wasted.

Cooperative management agreement = in the U.S. Fish and Wildlife Service, an agreement between the Service and another organization for cooperative management of fish, wildlife, and other species, and their requisite habitats.

Crenulation = in the context of a shoreline, having an irregularly wavy or serrated edge, margin, or outline.

Diaporiea = animal plankton found in cold deep lakes; important food source for many fishes.

Dichloro-diphenyl-dichloroethylene (DDE) = a chemical compound formed by the loss of hydrogen chloride (dehydrohalogenation) from the pesticide DDT; DDE is fat soluble which tends to build up in the fat of animals.

Dichloro-diphenyl-trichloroethane (DDT) = a widely used, very persistent pesticide in the chlorinated hydrocarbon group, now banned from production and use in many countries.

Earth Day = Earth Day is an annual event, celebrated on April 22, on which day events worldwide are held to demonstrate support for environmental protection.

Ecosystem = the interacting components of air, land, water, and all living things, including humans.

Ecosystem approach = an approach to management that accounts for the interrelationships among air, land, water, and all living things, including humans.

Effluent = wastewater discharged from industries or municipal wastewater treatment plants.

Enabling legislation = legislation that gives appropriated officials the authority to implement or enforce the law.

Endangered Species Act = a federal law in the U.S. that provides a program for the conservation of threatened and endangered plants and animals and the habitats in which they are found.

Environmentally sustainable economic development = economic development that does not degrade the environment nor natural resources; balanced and continuous economic, social, and environmental progress; development that meets the needs of the present generation without compromising the ability of future generations to meet their own needs; often called sustainable development.

Epidemic = a rapid spread or increase in the occurrence of a disease.

Eutrophication = the process of fertilization of aquatic ecosystems that causes high productivity and biomass; it can be a natural process or can be accelerated by human activities.

Evapotranspiration = loss of water from the soil both by evaporation from the soil surface and by transpiration from the leaves of the plants growing on it.

Exotic species = exotic species are those organisms introduced into habitats where they are not native.

Extirpated = local extinction of a species, although it still exists elsewhere.

Fashines = long bundles of live woody vegetation buried in a streambank in shallow trenches to help stabilize the bank, prevent erosion, and improve habitat.

Federal Water Pollution Control Act = initially signed in 1972 and amended numerous times in an effort to establish national programs intended to address all types of water pollution control programs; also called the Clean Water Act.

Fish tumor = an uncontrolled, abnormal, circumscribed growth of cells on or in a fish.

Fledge = stage in a young bird's life when the feathers and wing muscles are sufficiently developed for flight; the act of raising chicks to a fully grown state by the chick's parents to the point where they are able to fly.

Flyway = a route between breeding and wintering areas taken by migratory birds.

Food web = a network of food chains or feeding relationships by which energy and nutrients are passed on from one species of living organism to another.

Fur Trade Era = the time period in North American history (late-1600s to early-1800s) when fur trapping and trading were undertaken to meet European fashion demand.

Global warming = the continuing rise in the average temperature of the Earth's climate system.

Great Lakes Water Quality Agreement = a formal agreement between Canada and the United States to restore and maintain the chemical, physical and biological integrity of the Great Lakes Basin Ecosystem; it includes a number of objectives and guidelines to achieve these goals.

Green infrastructure = an approach that communities use to maintain healthy waters, provide multiple environmental benefits, and support sustainable communities; it uses vegetation, soils, and natural processes to manage water and create healthier urban environments.

Greenways = linear open spaces, including habitats and trails that link parks, nature reserves, cultural features, or historic sites for recreation and conservation purposes.

Governance = the system of rules, practices, and processes by which and institution or organization is directed and controlled; the act, process, or power of governing.

Habitat = it is a location where all attributes (i.e. physical, chemical and biological) occur to support a particular species; from a resource management perspective, it is the physical substrate that supports a biological community of organisms; for aquatic biota, it is typically depicted as three dimensional, including both the physical substrate and the overlying water.

Hand line trolling = a form of trolling in the Detroit River for Walleye using hand lines; hand-lining involves using minnows or minnow-like lures attached to a heavy weight placed close to bottom by a wire line; anglers hold the wire in their hand and bounce the weight off the river bottom, causing movement in the bait/lure that attracts Walleye.

Hard engineering = use of concrete breakwalls and steel sheet piling to stabilize the shoreline and protect developments from flooding and erosion, or to accommodate commercial navigation or industry; hard engineering provides no habitat for fish or other aquatic life.

Hydrologic pathway = various water movement pathways in the hydrologic cycle.

Hydrophytic = plants growing wholly or partially in water.

Impervious surface = constructed surfaces such as rooftops, sidewalks, roads, and parking lots that are covered by impenetrable materials such as asphalt, concrete, brick, and stone, and, therefore, repel water and prevent precipitation from infiltrating soils; soils compacted by urban development can also be highly impervious.

Important Bird Area = an area recognized as being globally important habitat for the conservation of bird populations; the program was developed by BirdLife Inernational and is administered in the U.S. by the National Audubon Society.

Indicator = a measurable feature that provides useful information on ecosystem status, quality, or trends, and the factors that affect them.

Industrial revolution = a rapid major change in an economy marked by the general introduction of power-driven machinery or by an important change in the prevailing types and methods of use of such machines.

International Joint Commission = a six-member commission established under the Boundary Waters Treaty of 1909 that assists Canada and the United States in finding solutions to problems in waters along the border.

International Migratory Bird Day = an annual celebration held each year during May that brings awareness on conserving migratory birds and their habitats throughout the Western Hemisphere.

Invasive species = an organism that causes ecological or economic harm in a new environment where it is not native.

Land Ethic = a philosophy that seeks to guide the stewardship of land when humans wish to make changes to it or use it (first defined by Aldo Leopold).

Landscape architect = a person involved in the planning, design, and sometimes direction of a landscape, garden, or distinct space.

Leadership in Energy and Environmental Design (LEED)-certified = an architectural standard for new buildings and ones being restored that recognizes best-in-class green building strategies and practices for sustainability.

Leptospirosis = an infectious disease of humans and of horses, dogs, swine, and other animals, caused by the spirochete *Leptospira interrogans* and characterized by fever, muscle pain, and jaundice, and in severe cases involving the liver and kidney.

Lithophilic = meaning "rock-loving"

Live stakes = a technique using live woody plant material for stabilizing streambanks and for repair of small earthen slips and slumps that are frequently wet; it is considered a relatively simple and inexpensive vegetative method for slope stabilization.

Loading = a unit describing the total weight of a substance carried at a given point in a river during a unit time (example - kilograms per day).

Macrobenthic invertebrates = invertebrate animals that are large enough to be seen with the naked eye and live at least part of the life cycle in the water.

Macrophytes = multicellular plant life found in aquatic ecosystems, normally visible to the naked eye.

Mangrove = any tropical tree or shrub of the genus *Rhizophora*, the species of which is mostly low trees growing in marshes or tidal shores, noted for its interlacing above-ground adventitious roots.

Marsh birds = marsh-dwelling water birds; water birds that require marsh habitats to meet life history requirements.

Master plan = a plan giving comprehensive guidance and instruction.

Mayflies = slender insects with delicate membranous wings having an aquatic larval stage and terrestrial adult stage usually lasting less than two days.

Memorandum of Understanding = a document that expresses mutual accord on an issue between two or more parties.

Mercury = a heavy metal that can be passed through the food chain and can concentrate in the fatty tissues of animals; it can be highly toxic and cause poisoning in humans.

Mesophytic = a classification of habitat where plant species are adapted to neither particularly dry nor particularly wet conditions; habitats where plants receive a well-balanced or moderate moisture supply.

Michigan Natural Features Inventory = a program in Michigan that utilizes teams of scientists with expertise in botany, zoology, aquatic ecology, and ecology to collect information about Michigan's native plants, animals, aquatic animals and natural ecosystems to aid in conservation of biodiversity.

Migratory Bird Conservation Commission = established in the United States in 1929, it was created to consider and approve any areas of land and/or water recommended by the Secretary of the Interior for purchase or rental by the U.S. Fish and Wildlife Service; it also is authorized to fix prices at which such areas may be purchased or rented.

Milligrams per liter (mg l^{-1}) = the most common unit of concentration used in water quality is milligram per liter (mg l^{-1}); it refers to a weight of substance in a volume of water.

Mixed use = development that combines two or more of the types of development, either residential, commercial, environmental, industrial, or institutional.

Municipal wastewater treatment = the process of removing contaminants from mixtures of human and other domestic wastes using physical, chemical, and biological processes.

Natural capital = the extension of the economic notion of capital (manufactured means of production) to environmental goods and services; the stock of natural ecosystems that yields a flow of valuable ecosystem goods or services into the future.

National Pollutant Discharge Elimination System (NPDES) = a U.S. regulatory system for controlling pollution of the environment; permits are issued that stipulate the quality of discharge and set time limits for compliance under Public Law 92-500 (i.e. commonly called the Clean Water Act).

National Wildlife Refuges Week = an annual week-long celebration of the National Wildlife Refuge System by the U.S. Fish and Wildlife Service and its many conservation partners.

Nature deficit disorder = a hypothesis by Richard Louv in his 2005 book titled *Last Child in the Woods* that human beings, especially children, are spending less time outdoors resulting in a wide range of behavioral problems.

Navigational channel = a river channel dredged and maintained to a specific depth to move ships and barges through to support industry and commerce.

Nonpoint source pollution = pollutants that are not discharged or emitted from a specific point source such as a pipe or smokestack; pollutants that enter the environment from diffuse sources such as surface runoff from precipitation.

Nursery grounds = nursery areas for juvenile fishes that are essential to their life cycles.

Ornithologist = a person who studies birds, including physiology, classification, ecology, and behavior.

Oxbow = a bow-shaped bend in a river, or the land embraced by it.

Paradigm shift = a change in basic assumptions within the ruling theory of science; acceptance by a majority of a changed belief, attitude, or way of doing things.

Phragmites = often called the common reed, is a large perennial grass found in wetlands throughout temperate and tropical regions of the world.

Phosphorus = an element that can affect water quality; in one of its forms, it can be used by algae in a stream or lake as a nutrient.

Place-based design = holistic design that seeks to ensure that what is being designed properly fits into the setting – its culture, history, surrounding land forms, existing ecosystem, etc.; it seeks to create a sense of place.

Point source = a discharge of wastewater from a fixed point such as a municipal wastewater treatment plant or industrial facility.

Pollution = making some resource less fit for some specific use.

Pollution sensitive organism = aquatic organisms that cannot live in polluted water; examples include mayflies, caddisflies, dragonflies, and others.

Polychlorinated biphenyls (PCBs) = a class of toxic organic compounds that are highly resistant to temperature and do not readily break down in the environment.

Primary treatment = the first stage of wastewater treatment, consisting of the removal of floating and settleable solids via sedimentation/settling.

Prohibition = the period of time from 1920-1933 in United States history when the manufacture, sale, and transportation of intoxicating liquors was outlawed.

Public Law 92-500 = an act passed by the U.S. Congress in 1972 that established national programs intended to control all types of water pollution; also called the Clean Water Act.

Punters = individuals who propel an open flat-bottom boat, usually with squared ends, in shallow waters using a long pole.

Ramsar Convention = an international treaty signed in 1971 for the conservation and sustainable use of wetlands.

Raptor = a bird of prey, such as a Hawk, Eagle, or Owl.

Refugia = plural of refugium; an area where special environmental circumstances have enabled a species or a community of species to survive after extinction in surrounding areas.

Remedial Action Plan (RAP) = a systematic and comprehensive plan to restore impaired beneficial uses in degraded areas of the Great Lakes using an ecosystem approach; outlined in the Canada-U.S. Great Lakes Water Quality Agreement.

Restrictive covenant = a provision in a deed limiting the use of property and prohibiting certain uses.

Return on investment = the concept of an investment of some resource yielding a benefit to the investor.

Riparian = relating to or living or located on the bank of a natural watercourse.

Riprap = rock material used to armor shorelines, streambeds, bridge abutments, pilings, and other shoreline structures, and protect them against scour, water, or ice erosion.

River = a natural stream of water of fairly large size flowing in a definite course or channel, or series of diverging and converging channels.

Rock revetment = in aquatic environments, these structures utilize rocks of varying sizes to absorb the energy of incoming water, thereby stabilizing the shoreline and protecting it from erosion.

Rust belt = the formerly dominant industrial region that is noted for the abandonment of factories, unemployment, outmigration, and overall decline.

Schematic plan = a scale drawing that outlines the floor plan where scale models can be placed for best and most effective positioning.

Secondary treatment = a type of treatment typically defined to be the best practical wastewater treatment for a municipal sewage treatment plant; conventional secondary treatment is achieved by screening, settling, and digestion of sewage by bacteria.

Sense of place = a characteristic held by people that make a place special or unique, that fosters a sense of authentic human attachment and belonging.

Sense of wonder = an intellectual and emotional state frequently invoked in discussions of science fiction. It is an emotional reaction to the reader suddenly confronting, understanding, or seeing a concept anew in the context of new information.

Shorebirds = birds that inhabits the open areas of beaches, grasslands, wetlands, and tundra; they include plovers, oystercatchers, avocets, stilts, and sandpipers; they often share characteristics of long legs, bills, and toes, and rather drab coloration.

Silent Spring = a book written by Rachel Carson in 1962 that awakened society to the effects of indiscriminate use of pesticides and helped start the modern environmental movement.

Soft shoreline engineering = use of ecological principles and practices to reduce erosion and achieve stability of shorelines and safety, while enhancing habitat, improving esthetics, and even saving money (sometimes called soil bioengineering).

Soil aggregation = the clumping or aggregation of soil particles that aids in infiltration of precipitation.

Soil bioengineering = a discipline of civil engineering that uses living plant materials to provide some engineering functions that increase shoreline stability and decrease erosion (sometimes called soft shoreline engineering).

Spawning grounds = a place where fish leave their eggs for fertilization.

Species at Risk = an extirpated, endangered, threatened species, or a species of special concern regulated by the Species at Risk Act in Canada.

Species diversity = a measurement of the number of species present in a given location or ecosystem.

Stakeholder = a general term used to describe any individual or organization who impacts or is impacted by a situation, resource, or process.

Stewardship = the careful and responsible management of something entrusted to one's care; in conservation, an ethic that embodies the responsible planning and management of natural resources.

Stonefly = any of numerous dull-colored primitive aquatic insects of the order Plecoptera, having a distinctive flattened body shape; a major food source for game fish like bass and trout; considered an indicator of good water quality and intolerant of water pollution.

Stopover sites = habitats that consistently provide migratory birds with the opportunity to refuel and rest during their journey.

Stressor = physical, chemical, or biological components of the environment that, when changed by human or other activities, can result in degradation of the environment or natural resources.

Subwatershed = a smaller basin within a larger drainage area where all of the surface water drains to a central point of the larger watershed.

Sustainability = the ability to maintain a certain status or process in existing systems; in relation to sustainable development it means the ability to meet the needs of the present generation without compromising the ability of future generations to meet their own needs.

Threatened species = any species (including animals, plants, fungi, etc.) which are vulnerable to endangerment in the near future.

Tipping point = in environmental science, it is the point in time where there is urgent need to take action and that, if nothing is done, society could see irreversible damage or harm to an ecosystem.

Total phosphorus = a measure of the total amount of inorganic and organic phosphorus.

Total suspended solids = the amount of filterable solids in a water sample; their presence in water reduces light penetration in the water column, can clog the gills of fish and invertebrates, and are often associated with toxic substances; in general, the higher the total suspended solids, the greater the water pollution.

Toxic substance = a substance that can cause death, disease, behavioral abnormalities, cancer, genetic mutations, physiological or reproductive malfunctions, or physical deformities in any organism or its offspring, or which can become poisonous after concentration in the food web or in combination with other substances.

Transboundary = referring to across or beyond boundaries.

Tubificid oligochaetes = aquatic sludge worms that are well recognized as pollution-tolerant because of their ability to thrive under poor water quality conditions or heavy pollution; their blood contains hemoglobin, which enables them to survive in waters where oxygen is lacking.

Typhoid fever = an infectious, often fatal, disease, usually contracted in the summer months, characterized by intestinal inflammation and ulceration, caused by the typhoid bacillus.

Underground Railroad = a network of secret routes and safe houses used by 19th Century slaves of African descent in the United States to escape to free states and Canada with the aid of abolitionists and allies who were sympathetic to their cause.

Uplands = areas of higher land; lands or an areas of lands that lie above the level where water flows or where flooding occurs.

Urban sprawl = the spreading of urban developments (as houses and shopping centers) on undeveloped land near a city.

Watershed = an area of land from which all water falling as rain or snow will flow toward a single point.

Wetland = an area of land whose soil is saturated with moisture either permanently or seasonally.

Wetland of International Importance = wetlands identified under Ramsar Convention as internationally important for the ecological functions and economic, cultural, scientific, and recreational values.

Wet Prairie = a native lowland grassland occurring on level, saturated, and/or seasonally inundated stream and river floodplains, lake margins, and isolated depressions.

Wildlife refuge = a geographical area where waters and lands are set aside to conserve fish, wildlife, and plants, but can also offer wildlife-compatible public uses like hunting, fishing, wildlife observation, photography, environmental education, and interpretation.

World Wetlands Day = an annual celebration of wetlands throughout the world that takes place on February 2nd each year under the auspices of the international Ramsar Convention.

Wyandot = indigenous peoples who live in North America and still live in the Detroit River watershed.

ECOVISION WORLD MONOGRAPH SERIES

The Ecovision World Monograph Series has been launched to focus on the paradigm of life in the aquatic ecosystems of our ever-changing planet and their sustainability under the impact of physical, chemical, biological, and human influences. It comprehensively covers the ecosystem science-based topics and themes pertaining to the ecology, health, biodiversity, and sustainability of whole ecosystems. The Ecovision Series is dedicated to integrated and ecosystemic research merging the high quality of a journal with the comprehensive approach of a book. The Series fosters trans-disciplinary and cross-sectoral linkages between the environment and ecological, socio-economic, political, ethical, legal, cultural, and management considerations.

Scope

- Ecovision publishes peer-reviewed compendiums of original papers on specific topics, and themes.
- It invites articles about the integrated assessment of aquatic environmental issues, including interactions with air, land, biota, and people.
- It promotes an integrated, multi-disciplinary, and multi-trophic approach.
- The Series focuses on the integrity, health, restoration, recovery, and management of stressed aquatic ecosystems.
- The Series explores the impact of environmental perturbations on ecosystem health at the structural and functional levels.

Themes

- State of aquatic ecosystems: freshwater and marine.
- Ecology, health, integrity, conservation, and management of ecosystems.
- Large ecosystems and Great Lakes of the World (GLOW).
- Ecology of coastal, harbour, delta, and other nearshore ecosystems.
- Food web dynamics and fisheries.
- Indicators of ecosystem health.
- Biodiversity, conservation, and sustainability.
- Impact of climatic events and climate change.
- Invasion of exotic species and their management.
- Effects of eutrophication, contaminants, and harmful algal blooms (HABs).
- Tools, techniques, models, and emerging technology.
- Adaptive management.
- Education, policy, and public outreach.

For more information please visit

www.aehms.org

Ecovision World Monograph Series

An international, integrated, peer reviewed scientific publication which explores the linkages between society, ecology, ecosystems and the environment.

List of Books Published:

1. Aquatic Ecosystems of China: Environmental and toxicological assessment, 1995.
2. The Contaminants in the Nordic Ecosystem: Dynamics, Processes, and Fate, 1995.
3. Bioindicators of Environmental Health, 1995
4. The Lake Huron Ecosystem: Ecology, Fisheries and the Management, 1995
5. Phytoplankton Dynamics in the North American Great Lakes, Vol. 1: Lakes Ontario, Erie and St. Clair, 1996
6. Developments and Progress in Sediment Quality Assessment: Rationale, Challenges, Techniques and Strategies, 1996
7. The Top of the World Environmental Research: Mount Everest - Himalayan Ecosystem, 1998
8. The State of Lake Erie Ecosystem (SOLE): Past Present and Future, 1999
9. Aquatic Restoration in Canada, 1999.
10. Aquatic Ecosystems of Mexico: Scope & Status, 2000.
11. Phytoplankton Dynamics in the North American Great Lakes, Vol. 2: Lakes Superior, Michigan, North Channel, Georgian Bay and Lake Huron, 2000.
12. The Great Lakes of the World (GLOW): Food-web, Health & Integrity, 2001.
13. Ecology, culture and conservation of a protected area: Fathom Five National Marine Park, Canada, 2001.
14. The Gulf Ecosystem: Health and Sustainability, 2002.
15. Sediment Quality Assessment and Management: Insight and Progress, 2003
16. State of Lake Ontario(SOLO): Past, Present and Future, 2003
17. State of Lake Michigan (SOLM): Ecology, Health and Management, 2005.
18. Ecotoxicological Testing of Marine and Freshwater Ecosystems: Emerging Techniques, Trends, and Strategies, 2005.
19. Checking the Pulse of Lake Erie, 2008.
20. State of Lake Superior, 2009.
21. Burning Rivers: Revival of Four Urban-Industrial Rivers that Caught on Fire, 2010

Dr. M. Munawar, Series Editor

Email: info@aehms.org
www.aehms.org

The Ecovision Series is dedicated to integrated and ecosystemic research merging the high quality of a journal with the comprehensive approach of a book.

DR. JOHN H. HARTIG –About the Author

Dr. John Hartig is trained as a limnologist with 30 years of experience in environmental science and natural resource management. He currently serves as Refuge Manager for the Detroit River International Wildlife Refuge and serves on the Detroit Riverfront Conservancy Board of Directors. From 1999 to 2004 he served as River Navigator for the Greater Detroit American Heritage River Initiative established by Presidential Executive Order. Prior to becoming River Navigator, he spent 12 years working for the International Joint Commission on the Canada-U.S. Great Lakes Water Quality Agreement. John has been an Adjunct Professor at Wayne State University where he taught Environmental Management and Sustainable Development, and has served as President of the International Association for Great Lakes Research. He has authored or co-authored over 100 publications on the Great Lakes, including *Bringing Conservation to Cities* and an earlier book titled *Burning Rivers* which was a 2011 Green Book Festival winner in the "scientific" category and a 2011 Next Generation Indie Book Awards finalist in the "science/nature/environment" category. John has received a number of awards for his work, including the 2013 Conservation Advocate of the Year Award from the Michigan League of Conservation Voters, the 2012 Outstanding Environmental Professional of the Year Award from the Michigan Association of Environmental Professionals, a 2010 Green Leaders Award from the Detroit Free Press, a 2005 White House Conference on Cooperative Conservation Award for Outstanding Leadership and Collaboration in the Great Lakes, the 2003 Anderson-Everett Award from the International Association for Great Lakes Research, and the 1993 Sustainable Development Award for Civic Leadership from Global Tomorrow Coalition.